Bio-Inspired Strategies for Modeling and Detection in Diabetes Mellitus Treatment

Bio-Inspired Strategies for Modeling and Detection in Diabetes Mellitus Treatment

Alma Y. Alanis
Division of Technologies for Cyber-Human Integration
CUCEI, University of Guadalajara
Guadalajara, Mexico

Oscar D. Sanchez
Division of Technologies for Cyber-Human Integration
CUCEI, University of Guadalajara
Guadalajara, Mexico

Alonso Vaca-Gonzalez
Institutional Safety, Health and Environment System (SISSMA)
University of Guadalajara
Guadalajara, Mexico

Marco A. Perez-Cisneros
Division of Technologies for Cyber-Human Integration
CUCEI, University of Guadalajara
Guadalajara, Mexico

MORGAN KAUFMANN PUBLISHERS

ELSEVIER AN IMPRINT OF ELSEVIER

ISBN: 978-0-443-22341-9

For information on all Morgan Kaufmann publications
visit our website at https://www.elsevier.com/books-and-journals

Publisher: Mica Haley
Acquisitions Editor: Chris Katsaropoulos
Editorial Project Manager: Palak Gupta
Production Project Manager: Selvaraj Raviraj
Cover Designer: Greg Harris

Typeset by VTeX

Working together
to grow libraries in
developing countries

www.elsevier.com • www.bookaid.org

Dedication

Alma Y. Alanis dedicates this book to her husband Gilberto, her daughters Alma Sofia and Daniela Monserrat, as well as her mother Yolanda for their unconditionally love. An acknowledgment goes to CUCEI of the University of Guadalajara for the support in this research.

Oscar D. Sanchez dedicates this book to his family, his parents Maria Vicenta and Arturo, his sisters Itzel and Sandra, to the help provided by his brother Arturo, he also wants to thank his uncle Ernesto and his partner Hannia. Also, special thanks go to the University of Guadalajara (CUCEI) for their support in this work.

Alonso Vaca-Gonzalez dedicates this book to his wife, who has provided her unconditional support throughout this process; to the research team of the Division of Technologies for Cyberhuman Integration of the University of Guadalajara, who gave him the confidence to participate in the team advising on the clinical aspects that feed the Artificial Intelligence models; and especially, to all the people living with diabetes and those without, who participated in the clinical examinations that gave rise to this great project.

Marco A. Perez-Cisneros dedicates this book to many generations of students that have helped him to enjoy teaching and doing research. Deep thanks to all.

Contents

List of figures

List of tables

Biographies

Alma Y. Alanis

She was born in Durango, Durango, Mexico, in 1980. She received the B.Sc degree from Instituto Tecnologico de Durango (ITD), Durango Campus, Durango, Durango, in 2002, the M.Sc. and the Ph.D. degrees in electrical engineering from the Advanced Studies and Research Center of the National Polytechnic Institute (CINVESTAV-IPN), Guadalajara Campus, Mexico, in 2004 and 2007, respectively. Since 2008 she has been with the University of Guadalajara, where she is currently a Chair Professor in the Department of Computer Science and from 2016 to 2019, she was dean of the PhD Program of Electronic and Computer Sciences. She is also a member of the Mexican National Research System (SNI-2) and a member of the Mexican Academy of Sciences. She has published papers in recognized International Journals and Conferences, and written five International Books. She is a Senior Member of the IEEE and Subject Editor of the Journal of Franklin Institute (Elsevier) and IEEE/ASME Transactions on Mechatronics, moreover she is currently serving on a number of IEEE and IFAC Conference Organizing Committees. In 2013, she received the grant for women in science by L'Oreal-UNESCO-AMC-CONACYT-CONALMEX. In 2015, she also received the Research Award Marcos Moshinsky. Since 2008 she has been a member for the Accredited Assessors record RCEA-CONACYT, evaluating a wide range of national research projects, in addition she has belonged to important project-evaluation committees of national and international research projects. Her research interests center on neural control, backstepping control, block control, and their applications to electrical machines, power systems, and robotics.

Oscar D. Sanchez

He was born in Texcoco, Mexico, Mexico in 1988. He received the B.Sc. degree in Computer Engineering in 2011 from the University of Guadalajara, the M.Sc. degree in Electronics and Computer Engineering, and the Ph.D. degree in Electronic and Computer Science from the University of Guadalajara in 2013 and 2018, respectively. Since 2015 he has been working at the Computer Department at the University Center for Exact Sciences and Engineering. His research interests are modeling and identification of systems, bioinformatics, optimization and prediction with intelligent systems.

Alonso Vaca-Gonzalez

He is a physician from the University of Guadalajara. He was born on December 14, 1989 in Guadalajara, Jalisco. He graduated with a Master's Degree in Medical Microbiology in 2021. He has various multidisciplinary diplomas in branches such as microbiology, hematology, public health, neurosciences, and occupational health. He was a cofounding member of the Early Detection and Guidance Program for Sexually Transmitted Infections (PRODOC-ITS). He carried out research activities in the private initiative evaluating new products and possible treatments for people with Diabetes Mellitus. His passion has led him to obtain the "Irene Robledo García" 2020–2022 Social Service Award for his outstanding work at the "Villa Primavera" Voluntary Isolation Center for COVID-19. Since 2016 he has been part of the Institutional System for Safety, Health and Environment (SISSMA) where he teaches human-capital training and skills-evaluation courses.

Marco A. Perez-Cisneros

Prof. Marco A. Perez-Cisneros received the B.S. degree with distinction in Electronics and Communications Engineering from the University of Guadalajara, Mexico, in 1995, the M.Sc. degree in Industrial Electronics from the ITESO University, Guadalajara, in 2000, and the Ph.D. degree from the Institute of Science and Technology, UMIST, The University of Manchester, UK, in 2004. Since 2005, he has been at the CUCEI campus of the University of Guadalajara, where he is currently the campus Rector. His current research interests include Computational Intelligence and Evolutionary Algorithms and their applications to Robotics, Computational Vision, and Automatic Control. He is a member of the Mexican Science Academy, the Mexican National Research System (SNI), and a member of the IEEE and the IET.

Preface

Diabetes mellitus is a metabolic disease characterized by high insulin levels caused by a defect in the secretion or action of the hormone insulin. This disease is prevalent in much of the world and poses challenges in terms of treatment and management. In these terms, emerging technologies in the area of Artificial Intelligence such as bio-inspired algorithms promise to be strategies that can revolutionize current therapies. This book presents approaches to some applications in medicine for the management of diabetes mellitus that are transforming the current landscape, such as parametric estimation of glucose–insulin dynamic models, advanced predictive models, and deep learning algorithms. A comprehensive analysis of how bio-inspired algorithms are optimizing diabetes prediction, diagnosis, treatment personalization, and ongoing management is presented. Machine learning applications, such as K-means, for the classification and diagnosis of diabetes are shown.

This work is the result of a collaborative effort between health experts and artificial intelligence professionals with the vision of integrating advanced strategies that are transforming the world in the treatment of diabetes mellitus.

This work is organized as follows:

• Chapter 1 presents, by way of introduction, the most important aspects of diabetes mellitus from a medical point of view, the diagnosis and treatment.

• Chapter 2 presents the current problems in the treatment and diagnosis of diabetes mellitus. In addition, two models are presented that describe the dynamics of glucose–insulin that are later used in Chapter 4. Also, the data used throughout the work is presented.

• Chapter 3 presents the fundamental concepts of this work, concepts that are used in Chapters 4, 5, 6, and 7.

• Chapter 4 presents the parametric estimation using evolutionary algorithms and algorithms based on particle swarms (presented in Chapter 3) of the mathematical models of Sorensen and Dalla Man using data from a patient with diabetes mellitus.

• Chapter 5 presents the identification of data obtained from a continuous glucose monitoring sensor using neural networks trained offline and networks trained online.

• Chapter 6 presents multistep prediction of blood glucose levels using neural networks. Two typical configurations were used: direct prediction and recursive prediction.

• Chapter 7 presents the classification and detection of diabetes mellitus and impaired glucose tolerance using deep neural networks. The classification is carried out on the real data described in Chapter 2.

• Chapter 8 presents the conclusions of this work.

Given the widespread applications in parametric estimation, modeling, prediction, classification, and diagnosis, it is difficult to limit the scope of this book to specific applications. This book is intended to function as a self-directed learning resource in addition to serving as a comprehensive textbook. The target audience is broad, it is useful for teachers, researchers, graduate students dedicated to artificial intelligence, data science, machine learning, modeling, estimation, identification, and classification.

Alma Y. Alanis
Oscar D. Sanchez
Alonso Vaca-Gonzalez
Marco A. Perez-Cisneros

Introduction

1.1 Diabetes Mellitus from a medical point of view

Diabetes Mellitus (DM) is a carbohydrate metabolic disorder condition that is characterized by high blood glucose levels; These elevated glycemias (blood sugar) lead to the inability or difficulty of the body to metabolize simple carbohydrates [1], [2]. This condition triggers a series of multiple macro- and micro-vascular effects in the human body, mainly damaging the cardiovascular system, the nervous system, the renal system, and even affecting the eyes, to the point of producing blindness, amputation of limbs, heart attacks, and even death [2], [3].

Depending on its etiology and pathophysiological mechanisms, DM is classified into different types: Diabetes Mellitus type 1 (DM1), Diabetes Mellitus type 2 (DM2), Gestational Diabetes (GD), and Diabetes Mellitus of Monogenic origin (MMD), which includes the MODY type (Maturity Onset Diabetes of the Young) [1], [4]. The etiological mechanisms of the different versions of DM are described in Table 1.1.

Table 1.1 Diabetes classification.

Type	Causes	Diagnostic tests
Type 1 or insulin dependent (DM1)	a) Autoimmune, by autoimmune destruction of β cells, usually leading to their absolute destruction and insulin deficiency. b) Idiopathic, due to unknown non-immunological mechanisms.	Glycemia, glucose tolerance curve, HbA1c
Type 2 (DM2)	Progressive defect in insulin secretion of a multifactorial nature.	Glycemia, glucose tolerance curve, HbA1c
Gestational diabetes (GDM)	Insulin resistance, Genetic predisposition.	Glucose tolerance curve, O'Sullivan Test
Monogenic diabetes	Point mutations in genes for β cell function, insulin, Genetic predisposition.	Specific genetic studies
Other types of diabetes	Exocrine diseases of the pancreas, endocrinopathies, induced by drugs or chemical agents, infections, other genetic syndromes and rare forms of immune-mediated diabetes.	Clinical criteria

According to the World Health Organization (WHO) [5], the prevalence of DM2 has increased considerably during the last three decades, with approximately 62 million cases in the Americas alone, while worldwide there are 422 million cases. Likewise, 244 084 deaths are attributed to it each year in the Americas alone, representing 1.5 million deaths worldwide, being the sixth leading cause of death worldwide. To maintain the health and quality of life of people living with diabetes, it is necessary to control diabetes, and even better, stop and even prevent its appearance in order to reduce this pandemic. The most frequent in our environment are DM1 and DM2; where DM1 occurs due to the lack of insulin production in the body by the beta cells of the pancreas; while DM2 is due to the cellular inability to introduce glucose into it, despite the existence of insulin [4], [6].

DM1 was colloquially known as juvenile or insulin-dependent diabetes, since it makes its debut in children, when its origin is genetic and because its treatment is entirely based on insulin; however, these features are not exclusive to it, so they are deprecated [1], [4]. DM2 is generally non-insulin dependent, since it is precipitated by a combination of many factors, mainly unhealthy dietary habits and a sedentary lifestyle, and there is no destruction of the pancreatic bed in the first instance, however, as it is a chronic and degenerative disease, the overproduction of insulin that the body performs in order to compensate for the lack of glucose uptake by the cells of the tissues, causing its wear and tear and, consequently, its failure, requiring the use of insulin [2], [7], [8]. It has also been described that the best treatment to prevent the complications of DM2 and reach health goals more easily is insulin-based therapy [2], [7], [8], [9].

For the onset of DM, a series of factors are involved, highlighting carbohydrate intolerance and a metabolic syndrome, which is derived from insulin resistance (IR) [4], [10], however, many factors are involved. Other factors, some are finer, such as the genetic predisposition given by HLA and non-HLA genes [11], and others more complex to characterize, such as hereditary-family history, hygienic-dietary habits, and even epigenetics [12].

When some of the adipose, muscle, and liver tissues do not respond adequately to the stimulus of insulin to be able to internalize glucose in their cells, an increase in the concentration of glucose in the blood occurs. This chronic exposure produces an ectopic storage of oxidizing energy substrates in the vascular bed, such as fatty acids and sorbitol, and therefore IR. The oxidant substrates will promote the inflammatory cascade mediated by cytokines and other precursors of metabolic stress, which will end up resolving in a lesion at the micro-vascular level. In late stages, up to 10 years or less of evolution, β cells will enter apoptosis, decreasing insulin production and further accentuating the body's inability to compensate for IR [2], [12], [13], [14].

1.2 Current diagnosis of Diabetes Mellitus and glucose intolerance

The diagnosis of DM is obtained by determining plasma glucose levels, either preprandially (fasting), postprandially 2 h after having consumed a 75 g glucose load orally, or by the criteria of glycosylated hemoglobin A1C (HbA1c). However, these separate tests may not detect abnormalities in the same individuals, see Table 1.2 [1], [4].

Before DM was established as a disease, a state in which there are high blood glucose levels without having diagnostic criteria for it having been documented, has been collo-

Table 1.2 Diagnosis criteria for Diabetes Mellitus.

Diagnostic tests	Diabetes Mellitus
Preprandial blood glucose	>126 mg/dl
Postprandial blood glucose*	> 200 mg/dl
HbA1c	> 6.5%
Random plasma glucose*	> 200 mg/dl

In the absence of unequivocal hyperglycemia, diagnosis requires two abnormal test results from the same sample or two separate test samples.

*Glucose tolerance test taken 2 h after oral load of 75 g of anhydrous glucose dissolved in water.

*As long as it presents classic symptoms of hyperglycemia or hyperglycemic crisis.

quially called "prediabetes" or glucose intolerance. These terms, together with the IR, have been used to classify people who have abnormal carbohydrate metabolism, with an upward trend in their figures, and therefore, a high probability of developing DM in the near future [1], [4], [6], [15]. Table 1.3 shows the diagnostic criteria for prediabetes.

Table 1.3 Comparison of diagnostic criteria for prediabetes and diabetes.

Diagnostic tests	Normal	Prediabetes		Diabetes Mellitus
		Altered fasting glucose	**Impaired glucose tolerance**	
Premeal blood glucose	<100 mg/dl	100–125 mg/dl	Does not apply	>126 mg/dl
Postprandial blood glucose*	<140 mg/dl	Does not apply	140–199 mg/dl	>200 mg/dl
HbA1c	<5–7%	5.7–6.4%		> 6.5%

Two determinations equal to or greater than the cut-off point of the same method or of two different ones are required for diagnosis.
*Glucose tolerance test taken 2 h after oral load of 75 g of anhydrous glucose dissolved in water.

Predicting the onset of DM has been an arduous and difficult task due to its multiple triggering factors, which is why different biomarkers have had to be monitored without much success, since most of them are produced and released in highly varied inflammation scenarios, regardless of high serum glucose levels or oxidizing agents such as sorbitol [13], [16], [17]. Being highly non-specific, it seems that continuous monitoring of the absorption and distribution of serum glucose, as well as the secretion of its related hormones, seems to be a better option to timely identify the onset of glucose-related disorders [1], [15].

Both IR and DM are conditions closely linked to the metabolic syndrome; even the IR has been raised as the cornerstone of the formation of the DM. IR is a biochemical–molecular concept, where there is less biological efficiency of insulin to internalize into cells due to multiple causes involving the same hormone or the behavior of its receptor or specific receptors, while a metabolic syndrome is a clinical concept characterized by the association of several pathophysiologically linked diseases through IR and hyperinsulinemia [1]. It has been suggested that at the onset of DM2, there is an alteration in insulin production that does not affect blood glucose (IR) levels, and that when this condition is maintained for a long time, the absorption of glucose in the tissues is decreased and ends up raising plasma glucose levels, which will result in Prediabetes or Diabetes, depending on its increase. However, it has not been possible to determine IR as the direct cause or consequence of DM, since it can occur without IR, the presence of IR does not exclude or confirm the presence of DM [6], [15], [18]. Therefore it is necessary to identify metabolic dysfunction changes from the beginning of their formation, even in states of Prediabetes.

Different methods have been described for the determination and diagnosis of IR, see Table 1.4, where the gold standard is the hyperinsulinemic clamp due to its greater sensitivity and specificity. This test consists of determining a variety of indices with the results obtained from the quantification of serum insulin peaks as glucose is administered parenterally in a controlled manner. However, being able to establish changes in the disorders of the levels of these components in the human body is extremely difficult, uncomfortable, and expensive for patients, due to the necessary follow-up, the available resources, and the multiple measurements and analyses that these entail [15]. Therefore having the ability to identify variations and trends with simple and inexpensive measurements, involving only the glycemic curve to stage and predict the behavior of carbohydrate metabolism will help the treating physician to timely address health problems before their appearance, complications or the disease per se, seeking the least number of samples taken. One of the tools that currently meet these requirements, because it is objective, easy to implement, and cheap, is the sequential quantification of glucose, basal and postprandial, with the glucose tolerance curve test [1], [15].

1.2.1 Oral glucose tolerance test

The oral glucose tolerance test (OGTT) is a test that consists of taking a single standardized glucose load and its subsequent monitoring of serum glucose in short periods of time, generally at 30 or 60 min until reaching 2 h, which over the years has been modifying its technique [19]. Currently, it is established that it should be performed in the morning with at least three days of free diet, that is, >150 g CHO, accompanied by physical activity. Fasting for at least 10 h but less than 16 h; patients can drink water. The patient must remain seated and smoking is prohibited during the entire test. A fasting blood sample is collected, at time zero a glucose dose of 75 g (1.75 g/kg of ideal body weight) is ingested in a concentration not greater than 25 g/dl of flavored water. Blood samples are collected at 30-min intervals for 2 h [4], [15], [20].

Conventionally, this test can classify patients into 3 stages: diabetic patient, patient with glucose intolerance or prediabetes, and healthy patient according to current ADA criteria, with the advantage of greater sensitivity when analyzing more than a single measurement, apart from the basal [1].

Thanks to the OGTT, it has been possible to define concepts that were previously ambiguous and that facilitate the integration of carbohydrate disorders in the body. Prediabetes is most often used to identify people whose glucose levels do not meet the criteria

Table 1.4   Diagnostic methods of insulin resistance.

Method		Advantages	Disadvantages
Direct	Euglycemic hyperinsulinemic clamp	Gold standard to assess sensitivity to insulin	Complex, invasive, difficult to perform in the pediatric population. They are not appropriate for use in large population studies or in routine clinical practice
	Hyperglycemic clamp	Gold standard for evaluating discharge of insulin	
	FSIVGT minimal model	Evaluate tissue sensitivity and secretion of insulin	
Indirect	Fasting plasma insulin	Methodologically simpler than direct	Variability according to pubertal development, poor correlation with the clamp
	HOMA index	Moderate to good correlation with the clamp	Highly variable cut-off points according to the population studied
	QUICKI Index		Breakpoints not available
	Matsuda–DeFronzo Index	Good correlation with the clamp	Multiple blood samples, IV catheter placement

HOMA, homeostasis model assessment; QUICKI, quantitative insulin sensitivity check index; FSIVGT, frequently sampled intravenous glucose tolerance test.

for diabetes, but have an abnormal carbohydrate metabolism, with an upward trend in their serum glucose levels with or without insulin production and who also have a high probability of developing DM in the near future [1], [4], [6], [15]; while glucose intolerance (IG) is the elevation of plasma glucose above 139 mg/dl, but below 200 mg/dl, after 2 h of taking 75 g of glucose in water [21]. However, we can also find the term impaired fasting glucose (AGA), defined as an 8-h fasting glucose concentration above 100 mg/dl but below 126 mg/dl [21]. Both GI and AGA are part of the diagnostic criteria for Prediabetes [1],12, [15], [21]. Table 1.5 shows the OGTT Diagnostic criteria.

Therefore the OGTT recognizes the alteration of the postprandial metabolism, making it a method capable of detecting diabetes more efficiently than other plasma glucose determinations. However, the variations in the glycemia curve and the few sporadic measurements decrease the sensitivity of the study and biases in the diagnosis appear, in addition to the fact that there is no standardized method for predicting complications of the disease by this method [1], [4], [6], [15]. For this, other laboratory techniques have been developed to facilitate its diagnosis, such as HbA1c, which provides an estimate of the average glycemia figures for approximately the last three months, which is a great advantage in the diagnosis of the disease with already established patterns. However, it provides us with little certainty and information to identify and prevent the development of diabetes or prediabetes, as well as in the prevention of diabetes complications [2], [4], [15]. For this,

Table 1.5   OGTT diagnostic criteria.

Diabetes Mellitus in nonpregnant adult		
Fasting value	**$\frac{1}{2}$-h, 1-h, or 1$\frac{1}{2}$-h OGTT**	**2-h OGTT**
venous plasma $\geq$ 140 mg/dl (7.8 mmol/L)		venous plasma $\geq$ 200 mg/dl (11.1 mmol/L)
venous whole blood $\geq$ 120 mg/dl (6.7 mmol/L)		venous whole blood $\geq$ 180 mg/dl (10.0 mmol/L)
capillary whole blood $\geq$ 120 mg/dl (6.7 mmol/L)		capillary whole blood $\geq$ 120 mg/dl (6.7 mmol/L)
Impaired Glucose Tolerance (IGT) in nonpregnant adults		
Fasting value	**$\frac{1}{2}$-h, 1-h, or 1$\frac{1}{2}$-h OGTT**	**2-h OGTT**
venous plasma $<$ 140 mg/dl (7.8 mmol/L)	venous plasma $\geq$ 200 mg/dl (11.1 mmol/L)	venous plasma of between 140 and 200 mg/dl (7.8 and 11.1 mmol/L)
venous whole blood $<$ 120 mg/dl (6.7 mmol/L)	venous whole blood $\geq$ 180 mg/dl (10.0 mmol/L)	venous whole blood of between 120 and 180 mg/dl (6.7 and 10.0 mmol/L)
capillary whole blood $<$ 120 mg/dl (6.7 mmol/L)	capillary whole blood $\geq$ 200 mg/dl (11.1 mmol/L)	capillary whole blood of between 140 and 200 mg/dl) (7.8 and 11.1 mmol/L)
Normal glucose levels in nonpregnant adults		
Fasting value	**$\frac{1}{2}$-h, 1-h, or 1$\frac{1}{2}$-h OGTT**	**2-h OGTT**
venous plasma $<$ 115 mg/dl (6.4 mmol/L)	venous plasma $<$ 200 mg/dl (11.1 mmol/L)	venous plasma $<$ 140 mg/dl (7.8 mmol/L)
venous whole blood $<$ 100 mg/dl (5.6 mmol/L)	venous whole blood $<$ 180 mg/dl (10.0 mmol/L)	venous whole blood $<$ 120 mg/dl (6.7 mmol/L)
capillary whole blood $<$ 100 mg/dl (100 mmol/L)	capillary whole blood $<$ 200 mg/dl (11.1 mmol/L)	capillary whole blood $<$ 140 mg/dl (7.8 mmol/L)

it is necessary to identify metabolic dysfunction changes from the beginning of their formation, such as IR, a condition linked to a metabolic syndrome. The first is a biochemical–molecular concept, where there is a lower biological efficiency of insulin to be internalized in cells due to multiple causes involving the same hormone or the behavior of its receptor or specific receptors, while the metabolic syndrome is a clinical concept characterized by the association of several pathophysiologically linked diseases through IR and hyperinsulinemia [1]. Being able to establish changes in the disorders of the levels of these hormones is extremely difficult and is uncomfortable and expensive for patients, therefore, having the ability to identify variations and trends only in the glycemic curve to stage and predict their behavior will help the doctor treating health problems in a timely manner before their complications or the disease itself appear with the least number of blood glucose measurements. This algorithm could benefit both hospital and home systems, implementing it in real-time quantification devices such as Abbott's LifeStyle (registered trademark) or even sophisticated software in hospitals and diagnostic laboratories.

1.3 Current treatment of Diabetes Mellitus

In order to understand the multiple treatments of the different types of Diabetes Mellitus, it is necessary to understand their pathophysiological causes to identify and solve the underlying metabolic problems.

We know that hyperglycemia is produced by three situations in particular: The decrease in insulin secretion due to an insulin-resistance effect with a decrease in peripheral glucose uptake at the muscle level and consequently the greater production of glucose at the hepatic level [6,21–23]. During the development of insulin resistance there is an increased insulin response, however, the function of pancreatic β cells is not normal, since it responds to the increase in blood glucose by increasing insulin secretion and therefore patients with intolerance to glucose may have elevated insulin levels. In addition, with the passage of time and the progression of the disease, the insular mass decreases and with it the production capacity of the pancreas to produce insulin, even in thin people living with DM2, they are highly resistant to insulin compared to their counterparts of control groups. It is estimated that once DM2 is diagnosed, the insular mass is around 50% and 5 years later, up to 75% is lost [6,23,24].

The causes of β-cell dysfunction are progression into old age, genetic factors such as genes linked to β-cell disruption such as TCFL2, increased insulin demand, autoimmune processes that attack the cell β, and the direct or indirect action of insulin resistance. Other causes have also been related, such as increased lipolysis, increased glucagon secretion, increased tubular reabsorption of glucose added to neurotransmitter dysfunction, abnormalities in the incretin axis, and even hypersecretion and islet amyloid polypeptide deposition putatively involved in progressive β-cell failure [25–27].

Each of these causes are targets in the control and treatment of Diabetes and are grouped as non-pharmacological and pharmacological, the latter is subdivided into insulin therapy and oral and injectable antidiabetics. It should be noted that one treatment does not exclude the other and combining them individually improves the probability of reaching the goals in less time, in order to follow a flexible treatment according to the needs of each patient and the treatment approach that needs to be emphasized [2,28,29]. In addition, the preferences and origin of the patient must always be taken into account in order to adapt any aspect of the treatment to their lifestyle and progress towards the evolution of healthier habits without interfering to a great extent with their beliefs and preferences [30].

General treatment guidelines are dietary recommendations, physical exercise, education for self-treatment, and drug treatment [9].

Adequate diet and constant physical activity are the key points of the treatment, with which control and/or remission of the disease can be achieved, in conjunction with pharmacological treatment. The diet of people living with diabetes includes carbohydrates, contrary to what is mistakenly believed; however, it must be modified by restricting, partially or totally, those of rapid action, which may or may not include any type of food as long as the portions are adequately counted [9]. The benefits of physical exercise lie in the reduction of insulin needs, it improves its sensitivity and reduces blood glucose levels. It should be done constantly and progressively, according to the physical abilities and condition of each person [9].

Regarding diabetes education for self-treatment, it must be carried out in a multidisciplinary way, consulting both the various medical specialists and the psychologist, nutritionist, physiotherapist, and physical trainer, in order to ensure that the patient is self-sufficient to be able to make decisions critically and effectively in various situations of daily life that manage to maintain good glycemic control [9].

To be considered an effective treatment, it is necessary to reach the therapeutic goals (see Table 1.6) by obtaining adequate individualized glycemic levels with various parameters focused on glycemic monitoring at different scales; Therefore it should be emphasized that, although the ideal is to maintain glycemic figures lower than those established in said goals, their maintenance is not completely strict, since their values can fluctuate between ranges depending on the priority needed in the face of a potentially associated to hypoglycemia event, disease duration, life expectancy, comorbidities, established vascular complications, patient preferences, and patient support system and resources [30].

Table 1.6   Glycemic control goals.

Criterion	Parameter
HbA1c	<7%
Preprandial blood glucose	80–130 mg/dl
Postprandial blood glucose	<180 mg/dl

HbA1c reflects the average serum glycemic level of the last 3 months derived from erythrocyte turnover, so it has a predictive value for diabetes complications and is therefore one of the parameters to be evaluated for good glycemic control. Its limitations lie in the slight variations depending on age, ethnicity, glycemic variability, kidney disease, anemia, or alterations in the erythrocyte turnover rate [30,31].

Capillary glycemic monitoring provides information regarding the current metabolic state, and gives us the guidelines to take the appropriate measures, both mediate and immediate, in the treatment and monitoring of the disease and direct efforts to address the identified weaknesses. It should be noted that it is essential in patients with DM1, however, it is not decisive in cases of DM2 [32]. It can be performed using different schemes, such as the one based on meals, in the detection of preprandial hyperglycemia or the detection of asymptomatic hypoglycemia [33,34]. The benefits of glycemic control objectives are listed in Table 1.7.

Continuous glycemic monitoring (CGM) is a recent tool that allows the identification of interstitial glucose levels through a special sensor, where two types of devices can be located: personal and professional. The former is intended for frequent and continuous use by the user and can measure and store blood glucose levels in real time continuously and without warning or intermittently, it being necessary to scan the device in order to store the values; while the professionals are from an institution that provides the service to the user, offering hidden or non-hidden data for certain and limited periods of time, around

Table 1.7 Objectives of capillary glycemic monitoring.

Specific objectives	Diabetes type 1	Diabetes type 2
Fasting, preprandial, and postprandial glycemic determination.	Yes	Yes
Insulin adjustment	Yes	Yes, in patients treated with insulin
Prevention of hypoglycemia	Yes	Yes, in patients treated with insulin or sulfonylureas
Intensification of glycemic control during physical activity, acute illness, etc.	Yes	Yes
Screening for unaware hypoglycemia	Yes, in patients with unrecognized hypoglycemia	No
Glycemic control where a sudden change in blood glucose is suspected	Yes	Yes

7 to 14 days, and whose objective is to evaluate glycemic patterns or trends. This resource has been shown to have an important predictive value for the reduction of HbA1c values in patients with DM1, while patients with DM2 have seen benefits both in treatments with or without basal insulin [30,35].

Pharmacological treatment, without insulin therapy, is aimed almost exclusively at patients with DM2, because they focus on the axes of its pathophysiological genesis: Greater peripheral glucose uptake, decreased insulin secretion, increased hepatic glucose production, increased lipolysis, reduced effect of incretins, increased glucagon secretion, increased tubular reabsorption of glucose, and neurotransmitter dysfunction [33].

The choice of pharmacological treatment should be based individually on both the intrinsic and extrinsic factors of each patient. These are understood as intrinsic to those conditioned by the progression of the diabetic disease and its complications, to the comorbidities present, the patient's own state of health, their weight, the risk of polypharmacy, contraindications, the efficacy of the medication, cardiovascular protection and renal, its safety and the tolerability, preferences, and the needs of each patient. The extrinsic ones are conditioned to aspects that do not depend on the patient, such as the costs and availability of the presentations, the routes of administration, the distribution and access to health services, the preference and knowledge of the treating physician, among others. All of the above apply to achieve adequate monitoring and control of the disease. The gradual introduction of drugs should be started, monitoring their tolerability and efficacy. It must be taken into account that medications that have not had an impact on glycemic control or weight should be discontinued, unless there is some added clinical benefit, such as cardiovascular or renal protection. If monotherapy does not reach the goal of HbA1c <7%, a healthy lifestyle, diet, and adherence to treatment should be reinforced, in addition to intensifying drug treatment and motivating the patient to reach therapeutic goals. As a rule, it is preferable to add drugs to the therapy and not interchange them, since it has been shown that the reinforcement of the pharmacological treatment increases the effectiveness of the reduction of HbA1c instead of substituting a drug in monotherapy, so it should not be to discontinue the medication unless it is intolerable or there are contraindications. In patients with DM2, insulin-based treatment is chosen when dual therapy fails to control HbA1c and metformin is not tolerated or is contraindicated, while in patients with DM1, insulin therapy is the standard treatment [33]. One of the most widely used approaches worldwide, considered a pillar of antidiabetic treatment, is metformin, a drug that improves insulin sensitivity through effects on its receptor signaling pathways and subsequent signaling cascade, due to its low costs, ease of treatment, few adverse effects, its safety as it does not produce hypoglycemia, it does not generate weight gain, there is a reduction of cardiovascular risk, and because it is one of the oldest drugs. In addition, it has been shown to reduce HbA1c by 1 to 2%, as well as the risk of death related to Diabetes complications. However, its main adverse effects are intestinal discomfort and diarrhea, which are not tolerated in 10% of cases; furthermore, it is contraindicated in patients with chronic kidney disease (CKD), notable liver disease, or another condition that compromises the tissue perfusion, so during its use renal function should be monitored [33].

On the other hand, we find the group of insulin secretagogues, which act directly on the pancreatic β cell to stimulate and increase insulin secretion. Within this group, we find the sulfonylureas (glibenclamide, glipizide, glicazide, glimepiride). Sulfonylureas are effective in reducing HbA1c by 1 to 2%, in addition to their low cost, extensive experience in their use, they are fast acting, once-daily administration, and glipizide, can be used during kidney dialysis, unlike its predecessors that are no longer used. Its main disadvantages lie in the risk of hypoglycemia, weight gain, dose adjustment required in case of renal dysfunction, and it has been seen that its prolonged exposure can desensitize β cells and reduce their secretory response, maintaining up to 25%, and decreasing its effectiveness even after 6 or 12 months of treatment. On the other hand, we have the meglitinides, secretagogues that manage to reduce HbA1c by around 1.57%. Thanks to their rapid onset and short duration action, patients require administration just 15 min before each meal, which lasts from 4 to 6 h, being appropriate management in patients with irregular meals and preprandial hyperglycemia, since they have a flexibility of dosage, although this implies the need for frequent daily doses. Its cost is moderate and it produces significant hypoglycemia, in addition to weight gain [33].

Thiazolinediones (pioglitazone, rosiglitazone) are peroxisome proliferator-activated receptor gamma (PPAR-γ) agonist drugs, which reduce hepatic glucose production (gluconeogenesis) and thereby modify the blood lipid profile and improve viability of the β cells. They reduce HbA1c by 0.7 to 1.6%, do not generate weight gain, reduce non-alcoholic steatohepatitis and cardiovascular risk, do not require dose adjustment in cases of kidney disease, they improve the lipid profile by increasing HDL cholesterol and decreasing triglycerides, and have a relatively low cost. In addition, they have a gradual onset effect to lower serum glucose levels, which do not generate hypoglycemia, but their characteristics vary depending on each individual and the factors that identify responders and non-responders are unknown. Among their adverse effects, there are the formation of edema, the decrease in bone mineral density and consequent osteoporosis and a greater risk of fractures, a greater risk of heart failure in predisposed patients, and they require renal monitoring and are contraindicated in people with bladder cancer [33].

During food digestion, nutrients in the food bolus induce the secretion of incretins, such as glucagon-like peptide-1 (GLP-1) and glucose-dependent insulinotropic polypeptide (GIP), formerly called gastric inhibitory polypeptide, which act in the pancreas increasing the secretion of insulin and decreasing the secretion of glucagon, a hormone that promotes hepatic gluconeogenesis, which releases glucose into the bloodstream. Both GLP-1 and GIP exert a feeling of satiety, promote gastric emptying, and are broken down by dipeptidyl peptidase 4 (DPP-4). GLP-1 receptor agonists (liraglutide, dulaglutide, semaglutide, lixisenatide) are drugs that try to mimic GLP-1 by activating its receptor, which induce insulin secretion induced by ingested nutrients in digestion. They have a certain versatility of administration, and their possible weekly administration can be considered an advantage. However, the half-life for some drugs in this group is only 2 to 4 h, as in the case of lixisenatide, so they must be administered daily and in conjunction with other drugs or insulin, being effective in lowering HbA1c both in monotherapy and combined in ranges by 0.6 to 1.8%, with the added effect of decreasing body weight by approximately 1 to 5 kg, reducing the cardiovascular risk and hospitalization for heart failure, they do not generate hypoglycemia, improve non-alcoholic steatohepatitis, and decrease albuminuria and postprandial hyperglycemia; on the other hand, their disadvantages are the same route of administration as injectable form, their high cost, their adverse effects, such as nausea and gastrointestinal discomfort, which self-limit in 4 to 8 weeks in most cases, which should not be administered in severe CKD, in patients with a history of pancreatitis or bile duct disease or medullary thyroid cancer. DPP-4 inhibitors (sitagliptin, vildagliptin, saxagliptin, linagliptin, tenegliptin) are drugs that increase the elevation of incretins by decreasing their degradation by DPP-4, decreasing their activity from 77% to 99%. They decrease HbA1c by 0.77%, do not cause weight gain or hypoglycemia, decrease postprandial hyperglycemia, lower blood pressure, are administered once daily (except vildagliptin) and can be used in patients with renal dysfunction (except linagliptin). Although their adverse effects are well tolerated, their hypoglycemic effect is modest, it is associated with a higher incidence of hospitalization for heart failure, it is not recommended in pancreatic disease, it can cause arthritis, bullous pemphigus, and decompensated heart failure, not to mention their high cost [33].

SGLT-2 inhibitors (dafaglifozin, empaglifozin, canaglifozin, ertuglifosin) are drugs that, by acting on this molecule at the renal level, reduce the absorption of glucose filtered by the kidney, promoting its renal excretion. They reduce HbA1c by 0.6 to 0.9%, cardiovascular risk, hospitalization for heart failure, and blood pressure, promote moderate weight loss, do not cause hypoglycemia, can be administered daily, and have nephroprotective properties. Their most frequent adverse effects are a predisposition to genitourinary infections, euglycemic ketoacidosis, a decrease in bone mineral density in the hip and, consequently, a greater risk of fracture in that area, an increase in LDL cholesterol, promotion of hypotension, they should not be used in ERC, and are the most expensive [33].

Insulin-based pharmacological treatment is focused on counteracting the insulin deficit present in patients with DM1 and other types of diabetes that present with pancreatic destruction. With insulin replacement, general metabolic control is sought. The only currently existing treatment alternative for people with DM1 is replacement with insulin. However, new technologies are still being tested aimed at halting the progression of progressive pancreatic destruction, when it is still identifiable; however, it is not possible to identify progressive damage to the pancreas without specific invasive studies. Other potential treatments under development for DM1 include pancreas–kidney and pancreatic islet transplantation, β-cell regeneration, and β-cell precursor stem cell implantation [33].

There are different types of insulins divided into two groups, prandial and basal, which are characterized based on their speed and time of biological activity, see Table 1.8, so they can be used according to dietary needs and the glycemic stability of each patient. The prandial ones require more rigorous control since they have a risk of hypoglycemia if they are not administered together with food, they are very useful to control hyperglycemia peaks, while the basal ones have the advantage of being able to be administered once a day, in addition to having less risk of producing hypoglycemia and favor less variability of glycemia peaks, since they lack maximum action peaks, being constant throughout their duration [33].

Table 1.8   Types of insulins.

Insulin	Type	Start of action	Maximum peak	Duration
Lispro, Aspart, Glulisine	Prandial, fast-acting analogs	5–15 min	45–75 min	2–4 h
Regular, Soluble, or Crystalline	Prandial, fast acting	30 min	2–4 min	5–6 h
NPH	Basal, intermediate action	$\approx$ 2 h	4–10 h	12–16 h
Detemir	Basal, intermediate action	$\approx$ 1–2 h	Does not apply	16–20 h
Glargina	Basal, long acting, and continuous	$\approx$ 1–2 h	Does not apply	20–24 h
Degludec	Basal, long acting, and continuous	$\approx$ 1–2 h	Does not apply	>24 h

Insulin replacement therapy aims to simulate, as far as possible, the physiological secretion of insulin under normal conditions, where constant insulin secretion occurs, to maintain blood glucose within normal limits between 80 to 100 mg/dl and also, in those situations in which food is consumed, where there is an increase in blood glucose and, consequently, an increase in pancreatic insulin secretion. Therefore an ideal treatment for DM1 is one that resembles this situation, see Fig. 1.1 [33].

Thanks to current insulin, it is increasingly easier to replicate endogenous insulin secretion than it was a few years ago. With the combination of a basal insulin, to keep blood glucose constant throughout the 24 h of the day, and a prandial insulin each time a meal is eaten to counteract the rise in blood glucose, that is, if the person with DM1 performs 3 food intakes, 3 doses of prandial insulin must be applied; if you make 5 intakes, it will be 5 doses. Therefore the insulin dose will be set, on the one hand, by the amount of food to be ingested, and on the other hand, the glycemic figures that it manages, adjusting according

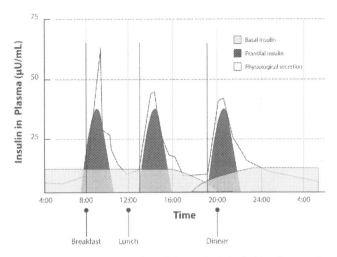

FIGURE 1.1 Purpose of insulinization. Under normal conditions, physiological insulin secretion shows increases proportional to serum glucose levels (yellow). The long and constant effect of basal insulins makes it possible to simulate the small permanent insulin concentrations throughout the day to regulate the states of gluconeogenesis and glycogenolysis in fasting periods (green), while prandial insulin, with an immediate and short effect, they will simulate the high peaks of insulin secretion during food intake to counteract the carbohydrate overload in the body (blue).

to the variability of these two factors. In general, the basal insulin dose is adjusted according to the fasting glycemic figures that the patient has, while the prandial one is based on the insulin prior to intake and 2 h after intake [33].

Insulin administration systems are a series of devices based on the subcutaneous administration of insulin by means of syringes and pens with fine needles, which are possible to manually adjust the dose and apply. Currently, it is possible to administer insulin with a continuous automated pump, both the setting and the application, eliminating needle sticks and greatly reducing dosing errors. Unfortunately, this technology is expensive and is not compatible with the characteristics of all patients [33,35].

1.4 Prediction of Diabetes Mellitus and its complications

Medical prevention is divided into three levels, where the primary one tries to prevent the disease from developing through patient education; the secondary focuses on the timely diagnosis of the disease, in its initial stages, to prevent complications from appearing and remit it as far as possible; Finally, tertiary prevention is responsible for stopping the progression of complications caused by the disease and providing treatment for them.

Preventing the development of DM then implies educating the vulnerable population, which presents intrinsic and extrinsic risk factors for its development, so that they avoid exposure to the extrinsic factors that induce its appearance, as well as taking measures for the intrinsic ones that may be modifiable. Preventing the complications of diabetes will require adequate control of the progression of the disease as well as continuous monitoring of the factors and markers of disease progression that show major metabolic alterations characteristic of micro- and macro-vascular lesions.

However, the prediction of diabetes behavior is even more complicated. We know from the natural history of the disease its clinical characteristics in the different stages and we can also estimate the time of progression of the established disease and monitor its complications, however, we have not been able to define the exact starting point of the disease. Also, due to the multi-factorial nature of diabetes, it has been impossible to accurately determine direct causative factors, and therefore to be able to identify the onset of the disease. Even in the case of DM1, which seems to be simpler than DM2, determining the onset of the disease is extremely complicated, since, contrary to popular belief, it has a complex pathogenesis.

Another example of the discrepancies between both diseases that hinder their prediction and complications is that of youth-onset diabetic kidney disease, which has a higher prevalence in people with DM2 compared to people with DM1. Furthermore, they have a more rapid progression towards albuminuria, they are more likely to reach the renal end-stage, and their mortality from all causes is higher. Regarding histopathological changes, we found that the glomerular tuft area, glomerular volume, mesangial matrix and its volume are increased in early-onset DM2 compared to what was found in DM1. Furthermore, the data obtained through sequencing identify transcriptional alterations that are shared between both types of diabetes and other specific alterations; however, the gene expression patterns of these alterations differ from each other, which has led to the need to maintain different treatment approaches. of kidney disease for both DM1 and DM2, even under the same pharmacotherapy, discrepancies in benefit and progression have been proven; where the renal benefits are clear in both types, but they are more pronounced in one type or the other depending on the pharmacological group used; however, many studies have scarce and contradictory data.

When the diagnosis of DM1 is made, due to the symptoms of hyperglycemia, the insular mass is generally at 20% to 30% of its capacity, being the latest phase of the disease. The natural history of the DM1 disease assigns three stages, where the first two are asymptomatic and the third symptomatic. The clinical characteristics to identify the first stage are the accumulation of at least two different autoantibodies directed to the insulin-producing pancreatic islets without changes in glycemic levels; as the damage to the insular mass progresses and insulin secretion decreases, subtle changes in blood glucose levels are obtained, without reaching the diagnostic criteria for diabetes, much less generating symptoms (stage 2); it is not until the almost total destruction of the insular mass occurs that the insulin produced is insufficient for the body's needs and symptoms of hyperglycemia occur, establishing the third stage [11].

Diabetes risk prediction models are based on multiple variables, such as patient characteristics (age, sex, ethnicity, weight, body mass index), medical history (comorbidities, time of disease, family history), blood parameters and urine, genetic biomarkers, diabetic control factors (diet, hypoglycemic drugs, insulin), among many others proposed; however, genetic, infectious, dietary, and humoral factors have prevailed as the most outstanding ones. In addition, these variables have ceased to be treated as individual, but rather as interactive factors that, when combined, would considerably increase the risk of developing the disease [11].

To date, the most effective way that has been proposed to prevent the development of DM2 is the reduction or limitation of its modifiable risk factors (overweight, obesity, poor diet, sedentary lifestyle, etc.) and carrying out regular monitoring or screening of people with non-modifiable factors (hereditary family history, high-risk ethnicities, history of cerebrovascular disease, high blood pressure, dyslipidemia, polycystic ovary, insulin resistance, history of gestational diabetes or prediabetes, HIV seropositive, age over 35 years) [1]; while monitoring the diabetic disease and its adequate control will avoid its complications.

All these variables are no longer treated as individual factors, but rather in an interactive way, which when combined, considerably increases the risk of developing the disease [11].

The FINDRISK questionnaire is designed to identify people at high risk of developing DM2 with the advantage of being able to be administered by health personnel or by oneself since it consists of 8 questions evaluated in a risk [1].

Several studies indicate that the CGM and the OGTT have an important role in predicting DM1 in people with risk factors, since they manage to identify constant patterns of hyperglycemia (>140 mg/dl), as well as a state of abnormal fasting and postprandial glycemic levels [36–38]. The OGTT has a better sensitivity for the diagnosis of DM1 and DM2 compared to HbA1c, since there is evidence of poor agreement between serum glycemia or the OGTT and HbA1c values (reference ADA 1). Furthermore, it is estimated that the inflection point of HbA1c <6.5% for a higher incidence of diabetic retinopathy could be poorly defined, which is why it has been proposed to lower it to <6.2% [39]. However, HbA1c has demonstrated great predictive capacity at 5 years.

The early detection of complications of diabetic disease for its consequent timely and immediate treatment, from the first signs and symptoms, will continue to be a fundamental pillar for prevention for a long time until new technologies can better predict the development of the illness. Recently, the use of artificial intelligence has been tested as a tool for the diagnosis, prediction, and control of various diseases. The use of Deep Learning, a state-of-the-art machine learning technique, has achieved good performance in the diagnosis of diabetic retinopathy, one of its most frequent complications, by using retinal image recognition, improving diagnostic accuracy, they reduce costs and increase access to procedures and specialty medical consultation [39].

1.5 Intelligent strategies inspired in biology and their applications in engineering and medicine

Let us start by defining what bio-inspired strategies are. Nature has generated biological systems from a process of evolution capable of overcoming real life problems. Nature-inspired computing or bio-inspired computing is an attempt to replicate the sophisticated biological mechanisms obtained from natural evolution such as robustness and adaptability. Natural Computing (NC) is a combination of the "inspired by nature" paradigm with "natural phenomena computing" and "using natural materials", which focuses on nature through computing. In general, Nature-inspired computing encompasses methods such as evolutionary computing, swarm intelligence, neural networks, and fuzzy logic. For its part, Computational Intelligence (CI) is closely related to NC. Computational Intelligence is defined as the application of strategies inspired by nature in real world problems.

Computational Intelligence encompasses the biology provided by different natural mechanisms that have been investigated and extensively studied, of which computational algorithms based on these mechanisms have been developed.

These mechanisms are, for example, the evolution that describes how living beings manage to adapt to their environment and improve after several generations, or mimicking brain functions to learn and remember, modeling the behavior of swarms like birds or fish or ants, even the reaction to intrusive pathogens in the body. These biological processes are modeled and coded in computational algorithms that are then adapted to specific application areas in the real world. In general, CI methods can be summarized as follows:

- Inspired by nature.
- Useful in real world problems.
- Solve complex problems without specific knowledge of the problem.

1.5.1 Main branches of Computational Intelligence

Computational Intelligence can be classified into the following main branches:

1.5.1.1 Evolutionary computation

Evolutionary Computation (EC) is a subset of Computational Intelligence inspired mainly by the biological process of evolution by simulating the process of natural selection. The main concepts of EC are:

- **Population:** This is the set of individuals, where each individual represents a possible solution to the problem.
- **Individual:** This is a single member of the population.
- **Aptitude:** This is a measure that is assigned to each individual and that indicates how well it performs with respect to the population in a given problem.
- **Generation:** This is an interaction in which a new population is generated by means of operators.
- **Selection:** This is the process of choosing individuals from the population to reproduce and pass on their characteristics to the next generation, generally based on their fitness.
- **Reproduction:** This is the process of generating new individuals according to the individuals selected through operators.
- **Mutation:** This is an operator to randomly change the genetic material of an individual.
- **Crossover:** This is an operator for the exchange of genetic information between two individuals to create new offspring.

- **Convergence:** This is the point at which the population has evolved over the generations to find a satisfactory solution to the problem.
- **Agent:** This is an individual particle that interacts with the environment in addition to other particles in the swarm.
- **Swarm:** A group of agents that interact with each other to accomplish a task.
- **Emergence:** The phenomenon in which complex behaviors or patterns arise from the interactions of simple agents.
- **Self-organization:** In this, the particles do not have a central authority, so the agents have the ability to coordinate their behavior among themselves.
- **Exploration:** This is the search for new solutions by trying different ideas, strategies, behaviors or actions.
- **Exploitation:** This is the use of ideas or strategies that have proven successful in the past to refine and improve solutions.
- **Fitness function:** This is a measure to evaluate the performance of the swarm.
- **Optimization:** The process of finding the best possible solution among a large number of candidate solutions.
- **Convergence:** The point at which the swarm has evolved to find a satisfactory solution to the problem.
- **Fitness Panorama:** This is the graph generated by the fitness of individuals in the population, showing the change in fitness across generations.
- **Optimization:** This is the process of finding the best possible solution among a large number of candidate solutions.

In general, evolutionary computation generates a population of possible solutions in the solution space, after applying operators such as mutation and crossover, new individuals are generated. After these individuals are evaluated according to their aptitude or how well they adapt or perform in the given problem, the most apt individuals are selected to transmit their characteristics to the next generation. The process is repeated for several generations until a satisfactory solution is found.

EC is a powerful tool in different fields of engineering, economics, medicine, among others, such as, for example, it is used in the field of robotics to generate control strategies for autonomous robots. It is useful in finance to optimize investment strategies and in biology to model the evolution of species. In engineering it has been used to design antennas, optimize energy systems, and control traffic flows. In addition, a promising area of research is in deep learning, combining neural networks with CE hybrid systems can be generated to solve more complex problems with more precise selection and faster convergence in neural network training.

1.5.1.2 Swarm intelligence

Swarm Intelligence (SI) imitates collective behaviors of social animals (ants, bees, and birds) to solve problems. Some of the SI concepts are:

- **Agent:** This is an individual particle that interacts with the environment in addition to other particles in the swarm.
- **Swarm:** A group of agents that interact with each other to accomplish a task.
- **Self-organization:** The particles do not have a central authority, so the agents have the ability to coordinate their behavior among themselves.
- **Exploration:** This is the search for new solutions by trying different ideas, strategies, behaviors or actions.
- **Exploitation:** This is the use of ideas or strategies that have proven successful in the past to refine and improve solutions.
- **Fitness function:** This is a measure to evaluate the performance of the swarm.
- **Optimization:** The process of finding the best possible solution among a large number of candidate solutions.
- **Convergence:** The point at which the swarm has evolved to find a satisfactory solution to the problem.

The main idea of SI is the use of simple agents that interact with their environment and with each other, due to their collaborative nature they achieve tasks that individual agents could not solve. Particle swarm-based algorithms typically employ a smaller particle population than an evolutionary algorithm. SI has topologies or sociometry to describe interconnections between particles so that each particle can be influenced by other agents, by switching rules, to move through the solution space. In the particle swarm the search ranges are scaled according to the consensus that the particles have with their neighborhood. The key advantage of SI is that it allows complex behavior and decision-making to emerge from a self-organizing, decentralized system, without the need for a central authority.

There have been a large number of research articles interested in SI, some of the applications are:

Combinatorial optimization, feature optimization, optimal path finding, image analysis, data mining, machine learning, bioinformatics, medical informatics, dynamical systems, industrial problems, finance, and business.

1.5.1.3 Neural network

A Neural Network (NN) or Artificial Neural Network (ANN) is a model inspired by studies of information processing in biological nervous systems, and the human brain. The main concepts of neural networks are:

- **Neurons:** These are the basic components of a neural network.
- **Layers:** These are sets of neurons and in general there are 3 types of layers, the input layer that receives input data, the output layer produces the final output of the network, and the hidden layer or layers that are between the input and output layers and serve to extract important features.

- **Weights:** These are used to determine the connection strength between two neurons. The weights are adjusted in the training process to minimize differences between the desired output and the network output.
- **Activation functions:** Activation functions are used to introduce nonlinearity into the output of a neuron.

In general, neural networks are made up of artificial neurons or nodes that communicate with each other. That is, nodes from other layers take information from other neurons, process the input, and produce an output that later serves as input to the neurons of the following layers. In general, the inputs of a neuron are multiplied by weights. Then, the multiplication of the weights by the inputs are added and evaluated in a mathematical function that determines the activation of the neuron. The result generates the output of the artificial neuron.

Neural networks have the ability to learn from examples, so they can be trained to recognize patterns, classify data, make predictions, or to solve particular problems by properly training specific examples. The training consists of adjusting the weights to obtain the desired output of the neural network. Typically, the weights are initialized in a reporting manner and adjusted until the error is reduced to the minimum possible.

Neural networks are quite popular due to their ability to handle large amounts of data and to form dense layers to learn increasingly complex patterns.

1.5.1.4 Fuzzy logic

Fuzzy Logic (FL) is conveniently useful for mapping an input space to an output space. FL is Precise Logic of Imprecision and Approximate Reasoning [40], which focuses on reasoning and decision making under uncertainty and ambiguity, and allows intermediate values ranging from completely true to completely false.

Fuzzy logic has the ability to handle imprecise and uncertain data, a very important feature in real-world applications, since most of the data is usually incomplete or ambiguous, so traditional methods do not work. In addition, FL is useful for manipulating nonlinear relationships between variables, so it allows modeling complex relationships that allow precise analysis [41]. The FL has applications in control, decision making, and artificial intelligence. Fuzzy logic concepts include:

- **Fuzzy sets:** These are used to represent imprecise or uncertain information. This set does not have a sharp and clearly defined limit, that is, it contains elements with degree and partial membership.
- **Membership function:** This is defined as a curve that assigns a membership value to each point in the input space (or degree of membership) between 0 and 1. They can be continuous or discrete, although it depends on the application.
- **Fuzzy rules:** These are a set of linguistic statements that relate inputs and outputs in a fuzzy system.
- **Fuzzy inference:** This is the process of mapping from a given input to an output using rules to make predictions or decisions.

- **Fuzzification:** This is the process of classifying numerical measurements into fuzzy sets.

1.5.1.5 Artificial immune system

Artificial Immune Systems (AIS) are computational models inspired by the biological immune system that respond to and develop immunity against foreign entities and are applied to problem solving.

The biological immune system consists of complex networks of cells, tissues, and organs that serve to defend the body from harmful entities. The immune system differentiates between its own entities and those of others, developing immunity by generating antibodies to specific antigens.

AIS is a set of algorithms to create a system that learns, recognizes, and responds to threats like a biological immune system does. The system can be trained on data to learn patterns and recognize anomalies, or run unattended to detect anomalies in the data [42].

These models have been applied to problems of intrusion detection, fault diagnosis, spam filtering, data classification, bioinformatics, and machine learning, such as artificial neural networks, etc. Artificial immune systems concepts include:

- **Recognition:** Defined as the ability to recognize, identify, and respond to a wide variety of different patterns, and to differentiate between own cells that malfunction and harmful foreign cells.
- **Feature extraction:** The ability of the immune system to extract features by filtering molecular noise from antigenic agents, before presenting them to other immune cells via antigen presenting cells (APC).
- **Diversity:** There are two main processes involved in generating diversity in the immune system. First, the recombination of gene segments to generate receptor molecules from gene libraries. This allows the immune system a great coverage of the universe of antigens. The second is somatic hypermutation, which helps with diversity in the immune system. Since immune cells reproduce in response to invading antigens, somatic mutation allows for the creation of new patterns of receptor molecules, to increase the diversity of immune receptors.
- **Learning:** Somatic hypermutation followed by a selection process allows the immune system to fine-tune its response to an invading pathogen; a process known as affinity maturation, which ensures that the immune system is becoming better and better.
- **Memory:** The process of maturation to the immune response allows cells and molecules that are successful in recognizing antigens to be maintained, in order to provide immune responses to future infections by the same or similar antigens. This is the principle behind medical vaccines.
- **Distributed detection:** Immune cells are specifically stimulated and respond to new antigens that may invade the body. Self-regulation: the population of the immune system is controlled by local interactions. That is, after an illness, the immune system reverts to a normal stable population until there is a need to respond to another antigen.

- **Metadynamics:** This is the continuous production, recruitment, and death of immune cells and molecules.
- **Immune network:** This proposes the immune system as a dynamic system whose cells and molecules are capable of recognizing each other, forming an internal communication network within the organism.

1.5.1.6 Medical applications using bio-inspired strategies

Among the medical applications in the area of evolutionary computing we can mention the evolutionary algorithm developed with a multi-objective purpose, which integrates Harris–Hawks decomposition and learning (MOEA/D-HHL) for medical machine learning in [43]. It was used to build machine learning algorithms for medical cancer gene expression data and clinical data for lupus nephritis and pulmonary hypertension. In the article [44], the sensitivity of biosensors was based on the evolutionary algorithm for biomedical applications. In another application, an extractive ensemble abstraction technique based on evolutionary algorithms was designed as a smart medical application with the idea of hybrid artificial intelligence in natural language processing [45]. For the segmentation of medical images, a clustering-based approach was used using a hierarchical evolutionary algorithm in [46].

Based on an adaptive barebones Salp swarm algorithm with quasi-oppositional learning, a Kernel Extreme Learning Machine model is improved to handle medical disease diagnostic problems [47]. In the paper [48] the stuck drilling mechanism is analyzed using the artificial fish swarm algorithm together with a support vector machine to generate a real-time early warning for the stuck medical drilling process. In the study published in [49], a method of synthesis of medical images for clinical diagnosis was proposed. The image enhancement Chameleon swarm algorithm was used that solves the problem of noise or low contrast of the input image. Similarly, in [50] Medical Image Segmentation was analyzed using an optimizer based on the reptile search algorithm that is combined with the Salp Swarm algorithm for image segmentation using grayscale multi-level thresholds.

Regarding medical applications, neural networks have been widely used, such as in [50], where the research of this work focused on specific medical predictions using a Naive Bayes classifier and artificial neural networks on the evolution of patients infected with the hepatitis B virus. Neural networks have also been used in early detection of skin cancer to prevent focal cell carcinoma and melanoma through image processing and machine vision. The convolutional neural network was used in [51]. In another research paper [52], an improvement in the learning of deep neural networks was proposed to help patients and physicians to improve the accuracy and reliability of the diagnosis and prognosis of cardiac diseases. It was based on a multilayer perceptron architecture with regularization and dropout through deep learning, for these tests a data set of 303 patients diagnosed with coronary disease was used.

Fuzzy Logic techniques can be used in pattern recognition with medical applications, as described in the [53] paper. In fact, in advanced precision medicine, genomic data is crucial to advances in cancer medicine. Therefore in [54], a FuzzyDeepCoxPH was proposed to identify high-risk missense mutation variants and candidate genes highly associated with cancer mortality.

In the work published in [55], a new method was proposed that combines a mathematical model that describes a chemotherapy treatment for breast cancer with an algorithm of the artificial immune system to increase the performance of predicting the size of a tumor. In another work [56] the artificial immune system method was implemented to increase the amount of data in the lung cancer data set and obtain a higher prediction rate for the diagnosis of lung cancer. Using an Artificial Immune System improves the accuracy of the abnormal distinctive pattern [57]. Obtaining the abnormal pattern by learning the characteristics of the entire data set.

In general, medical applications using bio-inspired strategies are varied and represent a cutting-edge line in innovation related to health, offering new alternatives in the treatment of diseases.

References

[1] American Diabetes Association Professional Practice Committee, 2. Classification and diagnosis of diabetes: Standards of medical care in diabetes – 2022, Diabetes Care 45 (Supplement_1) (2022) S17–S38.

[2] David M. Nathan, DCCT/Edic Research Group, The diabetes control and complications trial/epidemiology of diabetes interventions and complications study at 30 years: overview, Diabetes Care 37 (1) (2014) 9–16.

[3] Barbara K. Bailes, Diabetes mellitus and its chronic complications, AORN Journal 76 (2) (2002) 265–282.

[4] American Diabetes Association, et al., Office guide to diagnosis and classification of diabetes mellitus and other categories of glucose intolerance, Diabetes Care 14 (Supplement 2) (1991) 3–4.

[5] WHO, Global Report on Diabetes, World Health Organization, 2016.

[6] Felipe Pollak, Verónica Araya, Alejandra Lanas, Jorge Sapunar, Marco Arrese, Carmen Gloria Aylwin, Carmen Gloria Bezanilla, Elena Carrasco, Fernando Carrasco, Ethel Codner, et al., Il consenso de la Sociedad Chilena de Endocrinología y Diabetes sobre resistencia a la insulina, Revista Médica de Chile 143 (5) (2015) 627–636.

[7] Wendy L. Bennett, Lisa M. Wilson, Shari Bolen, Nisa Maruthur, Sonal Singh, Ranee Chatterjee, Spyridon S. Marinopoulos, Milo A. Puhan, Padmini Ranasinghe, Wanda K. Nicholson, et al., Oral diabetes medications for adults with type 2 diabetes: an update, 2011.

[8] Nuha A. ElSayed, Grazia Aleppo, Vanita R. Aroda, Raveendhara R. Bannuru, Florence M. Brown, Dennis Bruemmer, Billy S. Collins, Marisa E. Hilliard, Diana Isaacs, Eric L. Johnson, et al., 3. Prevention or delay of type 2 diabetes and associated comorbidities: Standards of care in diabetes – 2023, Diabetes Care 46 (Supplement_1) (2023) S41–S48.

[9] Boris Draznin, Vanita R. Aroda, George Bakris, Gretchen Benson, Florence M. Brown, R. Freeman, Jennifer Green, Elbert Huang, Diana Isaacs, Scott Kahan, et al., 5. facilitating behavior change and well-being to improve health outcomes: Standards of medical care in diabetes-2022, Diabetes Care 45 (Supplement_1) (2022) S60–S82.

[10] Omar Y. Bello-Chavolla, Rosalba Rojas-Martinez, Carlos A. Aguilar-Salinas, Mauricio Hernández-Avila, Epidemiology of diabetes mellitus in Mexico, Nutrition Reviews 75 (Suppl_1) (2017) 4–12.

[11] Marina Primavera, Cosimo Giannini, Francesco Chiarelli, Prediction and prevention of type 1 diabetes, Frontiers in Endocrinology 11 (2020) 248.

[12] Fermín I. Milagro-Yoldi, José A. Martinez, Epigenética en obesidad y diabetes tipo 2: papel de la nutrición, limitaciones y futuras aplicaciones, 2013.

[13] Kevin Luc, Agata Schramm-Luc, T.J. Guzik, T.P. Mikolajczyk, Oxidative stress and inflammatory markers in prediabetes and diabetes, Journal of Physiology and Pharmacology 70 (6) (2019) 809–824.

[14] John E. Gerich, Insulin-dependent diabetes mellitus: pathophysiology, in: Mayo Clinic Proceedings, vol. 61, Elsevier, 1986, pp. 787–791.

[15] Azucena Martínez-Basila, Jorge Maldonado-Hernández, Mardia López-Alarcón, Diagnostic methods of insulin resistance in a pediatric population, Boletín Médico del Hospital Infantil de México 68 (5) (2011) 397–404.

[16] Markku Laakso, Biomarkers for type 2 diabetes, Molecular Metabolism 27 (2019) S139–S146.

[17] Margarita Ortiz-Martínez, Mirna González-González, Alexandro J. Martagón, Victoria Hlavinka, Richard C. Willson, Marco Rito-Palomares, Recent developments in biomarkers for diagnosis and screening of type 2 diabetes mellitus, Current Diabetes Reports 22 (3) (2022) 95–115.

[18] Antonio González-Chávez, Mexican consensus of insulin resistance and metabolic syndrome, Revista Mexicana de Cardiología 10 (1) (1999) 3–19.

[19] J. Köbberling, D. Berninger, Natural history of glucose tolerance in relatives of diabetic patients: Low prognostic value of the oral glucose tolerance test, Diabetes Care 3 (1) (1980) 21–26.

[20] Feng Yu Kuo, Kai-Chun Cheng, Yingxiao Li, Juei-Tang Cheng, Oral glucose tolerance test in diabetes, the old method revisited, World Journal of Diabetes 12 (6) (2021) 786.

[21] Luis M. Romero-Mora, Felipe Durán-Íñiguez, Francisco J. Castro-Barajas, Hiperglucemia en ayuno e intolerancia a la glucosa el papel de los antecedentes familiares directos, Revista Médica del Instituto Mexicano del Seguro Social 51 (3) (2013) 308–312.

[22] Boris Draznin, Vanita R. Aroda, George Bakris, Gretchen Benson, Florence M. Brown, RaShaye Freeman, Jennifer Green, Elbert Huang, Diana Isaacs, Scott Kahan, et al., Summary of revisions: standards of medical care in diabetes-2022, Diabetes Care 45 (Suppl. 1) (2022) S4–S7.

[23] Ralph A. DeFronzo, Eleuterio Ferrannini, Donald C. Simonson, Fasting hyperglycemia in non-insulin-dependent diabetes mellitus: contributions of excessive hepatic glucose production and impaired tissue glucose uptake, Metabolism 38 (4) (1989) 387–395.

[24] Azucena Martínez-Basila, Jorge Maldonado-Hernández, Mardia López-Alarcón, Diagnostic methods of insulin resistance in a pediatric population, Boletín Médico del Hospital Infantil de México 68 (5) (2011) 397–404.

[25] Kenneth H. Johnson, Timothy D. O'Brien, Christer Betsholtz, Per Westermark, Islet amyloid, islet-amyloid polypeptide, and diabetes mellitus, New England Journal of Medicine 321 (8) (1989) 513–518.

[26] Juris J. Meier, Michael A. Nauck, Incretins and the development of type 2 diabetes, Current Diabetes Reports 6 (3) (2006) 194–201.

[27] Norma Urtiz-Estrada, Eduardo Lozano Guzmán, Olga D. López Guzmán, Ángel A. Hernández, Oscar Reza-García, Polimorfismos genéticos asociados a la diabetes mellitus tipo 2, Revista Mexicana de Ciencias Farmacéuticas 41 (4) (2010) 7–17.

[28] Diabetes Care, Addendum. 11. Chronic kidney disease and risk management: Standards of medical care in diabetes – 2022, Diabetes Care 45 (Suppl. 1) (2022) S175–S184.

[29] Alan J. Garber, Yehuda Handelsman, George Grunberger, Daniel Einhorn, Martin J. Abrahamson, Joshua I. Barzilay, Lawrence Blonde, Michael A. Bush, Ralph A. DeFronzo, Jeffrey R. Garber, et al., Consensus statement by the American Association of Clinical Endocrinologists and American College of Endocrinology on the comprehensive type 2 diabetes management algorithm – 2020 executive summary, Endocrine Practice 26 (1) (2020) 107–139.

[30] American Diabetes Association Professional Practice Committee, 6. Glycemic targets: standards of medical care in diabetes – 2022, Diabetes Care 45 (Supplement_1) (2022) S83–S96.

[31] David M. Nathan, Judith Kuenen, Rikke Borg, Hui Zheng, David Schoenfeld, Robert J. Heine, A1c-Derived Average Glucose (ADAG) Study Group, Translating the A1c assay into estimated average glucose values, Diabetes Care 31 (8) (2008) 1473–1478.

[32] Juan J. Gagliardino, Luiz Turatti, J. Davidson, J. Rosas-Guzmán, R. Castañeda-Limones, N. Ramos-Hernández, Manual de automonitoreo de la Asociación Latinoamericana de Diabetes (ALAD), La Revista de la Asociación Latinoamericana de Diabetes 18 (3) (2010) 120–126.

[33] Boris Draznin, Vanita R. Aroda, George Bakris, Gretchen Benson, Florence M. Brown, RaShaye Freeman, Jennifer Green, Elbert Huang, Diana Isaacs, Scott Kahan, et al., 9. Pharmacologic approaches to glycemic treatment: standards of medical care in diabetes – 2022, Diabetes Care 45 (Suppl. 1) (2022) S125–S143.

[34] Su Down, Nice type 2 diabetes management guidance: What's new?, Journal of Diabetes Nursing 26 (2) (2022).

[35] American Diabetes Association Professional Practice Committee, 7. Diabetes technology: standards of medical care in diabetes – 2022, Diabetes Care 45 (Supplement_1) (2022) S97–S112.

[36] Ping Xu, Craig A. Beam, David Cuthbertson, Jay M. Sosenko, Jay S. Skyler, Jeffrey P. Krischer, DPT-1 Study Group, Prognostic accuracy of immunologic and metabolic markers for type 1 diabetes in a high-risk population: receiver operating characteristic analysis, Diabetes Care 35 (10) (2012) 1975–1980.

[37] Andrea K. Steck, Fran Dong Iman Taki, Michelle Hoffman, Kimber Simmons, Brigitte I. Frohnert, Marian J. Rewers, Continuous glucose monitoring predicts progression to diabetes in autoantibody positive children, The Journal of Clinical Endocrinology and Metabolism 104 (8) (2019) 3337–3344.

[38] Andrea K. Steck, Fran Dong, Brigitte I. Frohnert, Kathleen Waugh, Michelle Hoffman, Jill M. Norris, Marian J. Rewers, Predicting progression to diabetes in islet autoantibody positive children, Journal of Autoimmunity 90 (2018) 59–63.

[39] Akihiro Nomura, Masahiro Noguchi, Mitsuhiro Kometani, Kenji Furukawa, Takashi Yoneda, Artificial intelligence in current diabetes management and prediction, Current Diabetes Reports 21 (12) (2021) 61.

[40] Lotfi A. Zadeh, Is there a need for fuzzy logic?, Information Sciences 178 (13) (2008) 2751–2779.

[41] M.R.h Mohd-Adnan, Arezoo Sarkheyli, Azlan Mohd-Zain, Habibollah Haron, Fuzzy logic for modeling machining process: a review, Artificial Intelligence Review 43 (2015) 345–379.

[42] Jon Timmis, Thomas Knight, Leandro N. de Castro, Emma Hart, An overview of artificial immune systems, in: Computation in Cells and Tissues: Perspectives and Tools of Thought, 2004, pp. 51–91.

[43] Mingjing Wang, Xiaoping Li, Long Chen, Huiling Chen, Medical machine learning based on multiobjective evolutionary algorithm using learning decomposition, Expert Systems with Applications 216 (2023) 119450.

[44] Irfan Ahmad Pindoo, Sanjeet K. Sinha, Increased sensitivity of biosensors using evolutionary algorithm for bio-medical applications, Radioelectronics and Communications Systems 63 (2020) 308–318.

[45] Chirantana Mallick, Asit Kumar Das, Janmenjoy Nayak, Danilo Pelusi, S. Vimal, Evolutionary algorithm based ensemble extractive summarization for developing smart medical system, Interdisciplinary Sciences: Computational Life Sciences 13 (2021) 229–259.

[46] Chih-Chin Lai, Chuan-Yu Chang, A hierarchical evolutionary algorithm for automatic medical image segmentation, Expert Systems with Applications 36 (1) (2009) 248–259.

[47] Jianfu Xia, Hongliang Zhang, Rizeng Li, Zhiyan Wang, Zhennao Cai, Zhiyang Gu, Huiling Chen, Zhifang Pan, Adaptive barebones salp swarm algorithm with quasi-oppositional learning for medical diagnosis systems: A comprehensive analysis, Journal of Bionics Engineering 19 (1) (2022) 240–256.

[48] Zhongyan Xian, Hai Yang, An early warning model for the stuck-in medical drilling process based on the artificial fish swarm algorithm and SVM, Distributed and Parallel Databases (2021) 1–18.

[49] Phu-Hung Dinh, Medical image fusion based on enhanced three-layer image decomposition and chameleon swarm algorithm, Biomedical Signal Processing and Control 84 (2023) 104740.

[50] Laith Abualigah, Mahmoud Habash, Essam Said Hanandeh, Ahmad MohdAziz Hussein, Mohammad Al Shinwan, Raed Abu Zitar, Heming Jia, Improved reptile search algorithm by salp swarm algorithm for medical image segmentation, Journal of Bionics Engineering (2023) 1–25.

[51] Ni Zhang, Yi-Xin Cai, Yong-Yong Wang, Yi-Tao Tian, Xiao-Li Wang, Benjamin Badami, Skin cancer diagnosis based on optimized convolutional neural network, Artificial Intelligence in Medicine 102 (2020) 101756.

[52] Kathleen H. Miao, Julia H. Miao, Coronary heart disease diagnosis using deep neural networks, International Journal of Advanced Computer Science and Applications 9 (10) (2018).

[53] Elena Vlamou, Basil Papadopoulos, Fuzzy logic systems and medical applications, AIMSN Neuroscience 6 (4) (2019) 266.

[54] Yang Cheng-Hong, Moi Sin-Hua, Hou Ming-Feng, Applications of deep learning and fuzzy systems to detect cancer mortality in next-generation genomic data, IEEE Transactions on Fuzzy Systems 29 (12) (2020) 3833–3844.

[55] OPhir Nave, Miriam Elbaz, Artificial immune system features added to breast cancer clinical data for machine learning (ML) applications, Biosystems 202 (2021) 104341.

[56] Melike Günay, Zeynep Orman, Tolga Ensari, Saliha Oukid, Nadjia Benblidia, Diagnosis of lung cancer using artificial immune system, in: 2019 Scientific Meeting on Electrical-Electronics & Biomedical Engineering and Computer Science (EBBT), IEEE, 2019, pp. 1–4.

[57] L. Sharmila, U. Sakthi, An artificial immune system-based algorithm for abnormal pattern in medical domain, Journal of Supercomputing 76 (2020) 4272–4286.

Problem statement

As discussed in the previous chapter, the current treatment for type 1 diabetes mellitus patients is based on periodic blood glucose measurements and insulin injections to regulate glycemic glucose values. However, regardless of whether it is DM1 or DM2, keeping glucose in normal ranges (between 70 and 180 mg/dl) is not a simple task, because there are alterations such as food intake, exercise, stress in addition to the complexity of glucose [1], [2].

Commonly, patients undergoing an exogenous insulin-based treatment can administer doses higher than necessary, so they experience hypoglycemia episodes (glucose levels below 70 mg/dl). Hyperglycemic and hypoglycemic risk scenarios cause macrovascular damage, such as peripheral, cerebral, and coronary insufficiency, is also associated with micro-vascular diseases such as neuropathy, retinopathy, and nephropathy.

In addition, if they extend for a long period of time, they cause a reduction in life expectancy, greater disability, which results in high costs in medical care for families [3].

Due to this, from an engineering point of view, new emerging technologies have developed precise sensors that continuously measure glucose, as well as devices for subcutaneous insulin infusion. These devices together enable the therapy for DM1 called a sensor-augmented pump (SAP) [4], which allows the patient to maintain glucose concentration in normal or glycemic ranges. Despite the great advances in SAP treatment, this is not automatic since the patient needs to calculate the amount of insulin needed according to their particular conditions at that time to program the insulin pump according to the measurements made by the sensor.

Automatic glucose control can be carried out by a self-contained biomedical insulin infusion device that has long been the main target of various research papers. This device is known as an artificial pancreas (AP) [5], [6], which aims to be minimally invasive to provide the best possible quality of life. A PA is a closed-loop system consisting of three main components, a subcutaneous insulin pump, a continuous glucose sensor (CGM), and a control algorithm that intelligently calculates insulin doses, a graphical representation can be seen in Fig. 2.1.

Of the three AP components mentioned above, the control algorithm is the most important since it is in charge of calculating the amount of insulin to be delivered to the subcutaneous tissue when there is food intake and postprandial periods. In the work published in [7], the PA was used continuously for a month, 24 h a day. A predictive control algorithm was used for this work. For its part, in [8] an individualized predictive control algorithm was proposed, whose novel proposal considers that the dynamics of insulin glucose varies between patients. That is why the effectiveness of the control algorithms depends to a great extent on the mathematical models that describe the dynamics of insulin glucose on which they were based.

Bio-Inspired Strategies for Modeling and Detection in Diabetes Mellitus Treatment
https://doi.org/10.1016/B978-0-44-322341-9.00011-2

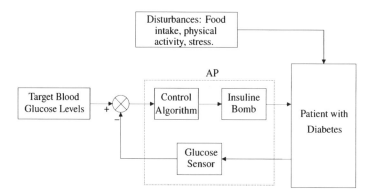

FIGURE 2.1 Representative diagram of the parametric estimation with evolutionary algorithms.

These biological models, in addition to being useful for the design of glucose controllers, have applications such as simulating different patient conditions or just testing glucose control algorithms before clinical trials, that is, to carry out in silico assessments. Some models represent the dynamics of diabetes mellitus from organs and tissues known as compartmental models, for example, the models of Sorensen and Dalla Man, these being mathematical models based on differential equations [9], [10]. Biological dynamics have also been modeled from neuronal networks, such as the multilayer perceptron neural network or, more recently, from a discreet neuronal network of high order. Regardless of the model used, its final application is the development of intelligent control capable of maintaining the patient's glucose in normal glucose ranges (approximately 70–120 mg/dl).

On the other hand, for the treatment of DM1 it is very beneficial to know future glucose conditions, especially risk scenarios such as hypoglycemia and hyperglycemia. Incorporating an alarm into the CMG in real time when glucose rises above the hyperglycemic threshold or falls below the hypoglycemic threshold could warn the patient or physician of these scenarios. However, it is preferable to anticipate hypoglycemic and hyperglycemic events before they occur [11], [12].

With adequate advance notice, these conditions could be prevented by significantly improving the therapy [13], [14]. Generating an alert 20 min to 30 min before the occurrence of risk events would provide enough information, either to the medical patient or to the glucose monitoring algorithm. Strategies capable of preventing future glucose concentration behaviors are necessary in DM [15]. With that knowledge, new glucose control algorithms can be developed to provide new approaches to glucose control by preventing hyperglycemia from low dose insulin or hypoglycemia from excessive insulin delivery.

In fact, there are some sensors that use simple methods to generate alerts when the glucose concentration trends such that the sensor believes the threshold has been crossed. Some methods that we can mention are [16], which is based on a polynomial, [17] uses the AR/ARMA model, state space in [18], the autoregressive (AR) model, autoregressive exogenous input (ARX), and autoregressive moving average exogenous input (ARMAX) [19], based on neural network models in [20], [21]. in the papers [22], [23], they propose a short-term glucose prediction, using weighted recursive least squares, the identification is performed online. According to the majority of published works, a prediction horizon between 30–45 min is sufficient to prevent risk scenarios in DM.

In addition, diabetes mellitus disease also presents problems in diagnosis in the early stages. Detection in early stages as well as correct diagnosis of any disease is as relevant as the treatment itself, since the patient has a better chance of improving with more and better information at his disposal [24]. Sometimes, the diagnoses are based on statistical and computational methods, because the data with which medical decisions are made consists of several heterogeneous variables that are obtained from different sources, such as demographic data, clinical history, medications, allergies, biomarkers, including medical photographs, markers or sensors. Each offers a different or partially different perspective on a patient's condition, the task that researchers and clinicians face when analyzing data such as dimensionality, the exponential increase in the number of samples in the feature space, and statistical features [25]. All of the above contribute to inaccurate diagnoses of diseases and delays, which implies inadequate patient care.

New approaches based on artificial intelligence (AI) for the systematic analysis of human diseases that allow a new vision that helps to detect aspects that are overlooked, have been developed [25]. As is the case of the work presented at [26], in which blood cancer is detected through images using convolutional neural networks, the detection of heart diseases [27] has also been carried out, Alzheimer's disease is diagnosed by a vector machine classifier [28], diabetes classification is carried out by machine learning methods using a high-dimensional medical data set [29]. In another article [30], machine learning models were used to classify glucose metabolism status in non-diabetics through an oral glucose tolerance test.

For their part, deep neural networks are a powerful tool in the area of machine learning that have allowed significant advances in different areas such as sequential data processing [31], sound and image processing [32], [33], in addition to medicine. In the field of medicine, studies have been reported on the diagnosis of diseases such as Parkinson's [34], heart failure in [35] or losses in the sensors of medical equipment [31]. In the case of this book, it is focused on the use of serial data obtained by sensors.

As can be seen, the challenges faced by diabetes mellitus are diverse and artificial intelligence, particularly bio-inspired strategies, can contribute to the development of new techniques for parametric estimation, identification, prediction, and detection.

2.1 State-of-the-art in modeling, identification, and detection of diabetes mellitus using bio-inspired strategies

In this section we present the state-of-the-art of some of the applications of bio-inspired algorithms for parametric estimation, identification, prediction, and detection.

2.1.1 Parametric estimation

Data modeling problems in finance, economics, medicine, among other sectors, present difficulties such as the lack of complete mathematical models of the data generation process, such as those available in physics, and the very limited amount of data available.

The construction of statistical models for this type of application, the selection and estimation of generalized models for prediction or classification are crucial and these models must be built in an optimal way, for example, the prediction and classification of risk scenarios of blood glucose levels in diabetic patients. Hence, when there are no nominal models of the data generation process, strategies such as non-parametric modeling are adopted. In this work, we address the two types of approaches for modeling type 1 diabetes mellitus.

In the case of diabetes mellitus, complex dynamic models have been developed to be used as diabetes simulators. Various types of models have been reported, among which is the simplest model of all, Bergman, and represents patients with T1DM in intensive care [36]. However, having low order does not include important physiological dynamics. For his part, Hovorka proposed a compartmental model, which consists of eight nonlinear differential equations that represent the subsystems of glucose, insulin, and insulin action [37]. The model proposed by Pedersen is based on the Hovorka [38] model of insulin absorption. Wilinska later developed a model to assess insulin administration in people with T1DM [39]. In addition, there are some other models that are more complex and descriptive of the biological mechanisms that interact in the glucose–insulin dynamics, for example, the model proposed by Tiran [40]. Later, based on Tiran's model, Guyton developed a model of glucose–insulin metabolism in normal individuals [41], which was later used by Cobelli from which is derived an integrated whole-body model for short-term glucose regulation. Sorensen relied on different characteristics of different models to present a compartmental model to represent organs and tissues that interact in glucose–insulin dynamics [9]. This model consists of 19 nonlinear differential equations made from a mass balance in each of the compartments. The organs that represent these compartments are brain, heart, lungs, intestine, liver, kidney, and periphery. Another model accepted as one of the most realistic models, was developed by Dalla Man, which describes the physiological behavior of a human being after food intake through seven compartments, glucose system, insulin system, gastrointestinal tract, muscle and adipose tissue, liver, and beta cells [42].

These models are useful to verify the efficiency of glucose controllers before clinical trials, or in silico evaluations. However, the development and validation of these models requires a lot of time, which implies a complete knowledge of physical phenomena. Of the previously mentioned models, the most used in the literature for the design of control algorithms and simulators for different scenarios of diabetes mellitus, we can mention the Sorensen, Dalla Man, and UVA/Padova models. To represent these scenarios, a critical issue is the parametric identification of a model with time series data for an individual or a population [43]. Frequently, the parametric characterization of the internal behavior of biological or physiological systems are not measurable [43]. Hence, the measurements are approximated indirectly as a parametric estimation problem. The parametric estimation can be possible through measurements of some of the state variables or output variables of the system. There are different methods that can be applied in the parametric estimation based on data observed by sensors. However, it is not an easy task since the physiological models contain numerous parameters and on some occasions many of them cannot be estimated [44]. In this book, we approach the use of bio-inspired algorithms to estimate parameters of representative models of glucose–insulin dynamics.

2.1.2 Neural identification

On the other hand, there are other alternatives to model glucose–insulin dynamics without complex medical knowledge. With essential knowledge of this disease, models can be obtained through stimulus–response methods using historical experimental data. Different works have made an effort to identify the dynamics of glucose–insulin. for example, using a linear approximation with ARX (AutoRegressive eXternal) [45], [19], ARMA (Automatic Moving Average) or ARMAX (Autoregressive Moving Average with exogenous inputs) models [46]. However, it is clear that linear models have limitations, so a better way of identifying dynamical systems is using nonlinear identification techniques, such as bio-inspired strategies, e.g., neural networks [47]. This methodology is widely used and is well established in engineering problems in modeling and control applications of nonlinear systems. The most widely used neural network architectures are recurrent and feedback [48], [49].

Neural networks for modeling the glucose–insulin dynamics of patients with T1DM is not a new task, in fact it has been considered in several works [47], [50], [51], [52], [53], [54]. More complex architectures were also used, such as neurofuzzy systems [50], [52].

Most of the reported neural models represent the metabolism of blood glucose and discard the purpose of control, that is, they do not obtain a model related to neural networks.

2.1.3 Prediction

In addition, several identified models also rule out multiple-step forward prediction. However, the models obtained using neural networks can also be used for the prediction of glycemic values several steps further.

In fact, based on continuous measurements from a glucose sensor, time series prediction methods can be implemented without the need for external inputs such as food intake, exercise, among others. Hence, it is possible to calculate the glucose trend without patient or doctor intervention.

Since the forecast error accumulates and propagates into the future very quickly, thanks to their nonlinear characteristics, neural networks are an alternative to predicting glucose levels using data from glucose sensors.

Works using neural networks for glucose prediction are mentioned below. In [20] a neural model was proposed that used as inputs measurements made by a continuous glucose monitoring sensor to predict glucose levels at 15, 30, and 45 min ahead, the training was performed offline. Similarly, in [55], measurements from a glucose sensor were used to predict future glycemic values. The nonlinear autoregressive neural network (NAR) and the neural network with long and short-term memory (LSTM) were used with a prediction horizon of 30, 45, 60, and 90 min. In an interesting study [56], a predictor made up of two parts was proposed, a model to describe the linear dynamics and a neural network that corrects the error made by the linear model. In addition to using measurements from a continuous glucose sensor, it also incorporated information such as food intake to predict 30 min ahead. In another work [57], the LSTM, BiLSTM, Convolutional LSTM and Time Convolutional neural networks were used for the prediction of glucose 30 and 60 min ahead.

2.1.4 Detection

Finally, a recent little-exploited application in medicine is the classification and detection of diseases in early stages. Thanks to new advances in machine learning, they have allowed early detection and diagnosis of DM through automated processes. Papers have recently been published about the detection and diagnosis of DM using methods of machine learning and artificial intelligence. For example, in [58], feature extraction and classification using a CNN and Support Vector Machine from retinal photos, having a precision of 95% were suggested. DM1 requires an exogenous supply of insulin to control blood glucose. The method proposed by [59] used the K-Nearest Neighbor Classifier to estimate the postprandial glucose profile for recommended time and adjustment of the food bolus. This method was tested for adults in silico with the UVA/PADOVA system and approved by the Food and Drug Administration as a supplement for animal research. In [60] the prediction of MD based on iridology or based on iris was developed using ML algorithms. ML was used to automate the discovery process. KNN was used to distinguish healthy patients and diabetic groups. The results were found to have an accuracy of 85.6%, the false positive rate was 11.07%, the false negative rate was 20.40%, the specificity was 0.889 and the sensitivity was 0.796. Deep learning methods extract features from large amounts of data and learn these features indirectly when trained, allowing it to differentiate between the volume of DM information to deal with intraclass differences and noisy information that is present in the DM data. Some Deep learning models that were used in DM detection are MLP [61], Feedforward Neural Network [62], [63], Deep CNN [64], [65], [66], [67], [68], and Extreme Learning Machine [69], [70], [71], [72].

2.2 Compartmental models Sorensen and Dalla Man

Compartmental models that describe the dynamics of DM1 have been a powerful tool for understanding, predicting, and designing controllers. The compartmental models are generally composed of sub-models that describe the main processes that affect glucose metabolism in patients with diabetes. Some common compartments are organs or tissues. These models contain numerous parameters that must be estimated, but once correctly fitted to data from patients with diabetes they can be simulated to examine different diabetic conditions or the effects of control actions. Close to reality behavior will depend on each setting and can be quantified in terms of a performance measure that indicates the benefit or harm associated with a particular therapy. Then, alternative therapies can be evaluated by comparing their respective measures of performance. Conceptually, compartmental models appear to be attractive, which is why compartmental models of glucose–insulin dynamics have been proposed for potential use in clinical therapy planning in diabetes care, such as the widely used models of Sorensen and Dalla Man. In this section, we present the compartmental models used to perform the parametric estimation task.

2.2.1 Sorensen model

The model proposed by Sorensen is a system consisting of 19 nonlinear equations. Through three subsystems the dynamics of glucose are represented: insulin, glucagon, and metabolic rates. The detailed description of the model can be found in [9]. Hence, to reproduce the dynamics of DM1, the following assumptions are made; first the release of insulin by the pancreas is omitted and secondly the scale of metabolic functions can be changed in such a way that the blood glucose levels correspond to a patient with DM1 concerning parametric identification. The set of equations of the Sorensen model are shown in Section 2.2.2. In addition, for greater clarity, the equations are grouped into the three subsystems to which they belong. In Appendix A.2 the set of nominal values of the Sorensen parameters are included.

2.2.2 Glucose subsystem

The first subsystem that we show is the one corresponding to glucose that involves six compartments (brain and central nervous system; heart and lungs; adipose tissue and skeletal muscle; stomach and small intestine; liver and kidney). The compartments connect directly. The compartments are a minimal set of physiological processes to isolate the unit of glucose metabolism in said organs and tissues. As a result, the glucose subsystem has eight differential equations with nonlinear terms:

Brain (vascular tissue):

$$\frac{dG_{BV}}{dt} = \frac{Q_B^G}{V_{BV}^G}(G_H - G_{BV}) - \frac{V_{BI}}{V_{BV}^G T_B}(G_{BV} - G_{BI}) \tag{2.1}$$

Brain (interstitial tissue):

$$\frac{dG_{BI}}{dt} = \frac{V_{BI}}{V_{BI} T_B}(G_{BV} - G_{BI}) - \frac{\Gamma_{BGU}}{V_{BI}} \tag{2.2}$$

Heart and lungs:

$$\frac{dG_H}{dt} = \frac{1}{V_H^G}(Q_B^G G_{BV} + \left(Q_L^G G_L + Q_K^G G_K + Q_P^G G_{PV} - Q_H^G G_H - \Gamma_{RBCU}\right)) \tag{2.3}$$

Gut:

$$\frac{dG_G}{dt} = \frac{Q_G^G}{V_G^G}(G_H - G_G) + \frac{1}{V_G^G}(\Gamma_{meal} - \Gamma_{GGU}) \tag{2.4}$$

Liver:

$$\frac{dG_L}{dt} = \frac{1}{V_L^G}(Q_A^G G_H + Q_G^G G_G - Q_L^G G_L + \Gamma_{HGP} - \Gamma_{HGU}) \tag{2.5}$$

Kidney:

$$\frac{dG_K}{dt} = \frac{Q_K^G}{V_K^G}(G_H - G_K) - \frac{\Gamma_{KGE}}{V_K^G} \tag{2.6}$$

Periphery (vascular tissue):

$$\frac{dG_{PV}}{dt} = \frac{Q_P^G}{V_{PV}^G}(G_H - G_{PV}) - \frac{V_{PI}}{T_P^G V_{PV}^G}(G_{PV} - G_{PI}) \tag{2.7}$$

Periphery (interstitial tissue):

$$\frac{dG_{PI}}{dt} = \frac{V_{PI}}{T_P^G V_{PI}}(G_{PV} - G_{PI}) - \frac{\Gamma_{PGU}}{V_{PI}} \tag{2.8}$$

2.2.3 Insulin subsystem

The dynamics of insulin are similar to physiological glucose, with the only difference that the insulin subsystem considers the pancreas as one more compartment. However, this compartment is eliminated according to the aforementioned assumption for patients with DM1. In addition, the dynamics of insulin in the interstitial fluid of the brain was not considered since the brain cell membrane is impermeable to the passage of insulin in the cerebrospinal fluid [37]:

Brain (vascular tissue):

$$\frac{dI_B}{dt} = \frac{Q_B^I}{V_B^I}(I_H - I_B) \tag{2.9}$$

Heart and lungs:

$$\frac{dI_H}{dt} = \frac{1}{V_H^I}(Q_B^I I_B + Q_L^I I_L + Q_K^I I_K + Q_P^I I_{PV} - Q_H^I I_H + i(t)) \tag{2.10}$$

Gut:

$$\frac{dI_G}{dt} = \frac{Q_G^I}{V_G^I}(I_H - I_G) \tag{2.11}$$

Liver:

$$\frac{dI_L}{dt} = \frac{1}{V_L^I}(Q_A^I I_H + Q_G^I I_G - Q_L^I I_L + \Gamma_{PIR} - \Gamma_{LIC}) \tag{2.12}$$

Kidney:

$$\frac{dI_K}{dt} = \frac{Q_K^I}{V_K^I}(I_H - I_K) - \frac{\Gamma_{KIC}}{V_K^I} \tag{2.13}$$

Periphery (vascular tissue):

$$\frac{dI_{PV}}{dt} = \frac{Q_P^I}{V_{PV}^I}(I_H - I_{PV}) - \frac{V_{PI}}{T_P^I V_{PV}^I}(I_{PV} - I_{PI}) \tag{2.14}$$

Periphery (interstitial tissue):

$$\frac{dI_{PI}}{dt} = \frac{V_{PI}}{T_P^I V_{PI}^I}(I_{PV} - I_{PI}) - \frac{\Gamma_{PIC}}{V_{PI}} \tag{2.15}$$

2.2.4 Metabolic rates and dynamics of glucagon

Finally, the metabolic rates that contribute to the mass balance describe the physiological process in some of the compartments. In the case of the glucose subsystem, seven metabolic rates are considered, three for the insulin subsystem, and finally two each for the plasmatic glucagon elimination rate and the plasmatic glucagon release rate.

Brain glucose uptake rate:

$$\Gamma_{BGU} = 70\,\text{mg/min} \tag{2.16}$$

Red blood cell glucose rate:

$$\Gamma_{RBCU} = 10\,\text{mg/min} \tag{2.17}$$

Glut glucose uptake rate:

$$\Gamma_{GGU} = 20\,\text{mg/min} \tag{2.18}$$

Periphery glucose uptake rate:

$$\Gamma_{PGU} = 35\frac{G_{PI}}{86.81}\{7.03 + 6.52tanh[0.388(\frac{I_{PI}}{5.304} - 5.82)]\} \tag{2.19}$$

Hepatic glucose production rate:

$$\Gamma_{HGP} = 155M_{HGP}^I\{2.7tanh(0.39G_C) - f_2\}\left\{1.42 - 1.41tanh\left[0.62\left(\frac{G_L}{101} - 0.497\right)\right]\right\} \tag{2.20}$$

Hepatic glucose production mediated by insulin:

$$\frac{d}{dt}M_{HGP}^{I} = \frac{1}{25}\left\{1.21 - 1.14tanh\left[2.44\left(\frac{I_L}{21.43} - 0.89\right) - M_{HGP}^{I}\right]\right\} \tag{2.21}$$

$$\frac{df_2}{dt} = \frac{1}{65}\left[\frac{2.7tanh(0.39G_C) - 1}{2} - f_2\right] \tag{2.22}$$

Hepatic glucose uptake rate:

$$\Gamma_{HGU} = 20M_{HGU}^{I}\left\{5.66 + 5.66tanh\left[2.44\left(\frac{G_L}{101} - 1.48\right)\right]\right\} \tag{2.23}$$

Solution of hepatic glucose uptake mediated by insulin:

$$\frac{d}{dt}(M_{HGU})^{I} = \frac{1}{25}\left[2tanh\left(0.55\frac{I_L}{21.43} - M_{HGU}^{I}\right)\right] \tag{2.24}$$

Kidney glucose excretion rate:

$$\Gamma_{KGE} = \begin{cases} 71 + 71tanh[0.11(G_K - 460)], & 0 \le G_K < 460\,\text{mg/min} \\ -330 + 0.872G_k, & G_K \ge 460\,\text{mg/min} \end{cases} \tag{2.25}$$

Hepatic insulin clearance rate:

$$\Gamma_{LIC} = F_{LIC}[Q_A^I I_H + Q_G^I I_G + \Gamma_{PIR}] \tag{2.26}$$

$$\Gamma_{LIC} = 0.40 \tag{2.27}$$

Pancreatic insulin release rate:

$$\Gamma_{PIR} = 0.0 \tag{2.28}$$

Kidney insulin clearance rate:

$$\Gamma_{KIC} = F_{KIC}[Q_K^I I_K] \tag{2.29}$$

$$\Gamma_{KIC} = 0.30 \tag{2.30}$$

Periphery insulin clearance rate:

$$\Gamma_{PIC} = \frac{I_{PI}}{\left[\left(\frac{1 - F_{PIC}}{F_{PIC}}\right)\left(\frac{1}{Q_P^I}\right) - \left(\frac{T_P^I}{V_{PI}}\right)\right]} \tag{2.31}$$

$$\Gamma_{PIC} = 0.15 \tag{2.32}$$

Plasmatic glucagon release rate:

$$\Gamma_{PTR} = 2.93 - 210 tanh\left[4.18\left(\frac{G_H}{91.89} - 0.61\right)\right]1.31 - 0.61 tanh\left[1.06\left(\frac{I_H}{15.15 - 0.47}\right)\right] \quad (2.33)$$

Only one compartment was used for modeling the counter-regulatory effect of glucagon on the glucose–insulin system.

$$\frac{d}{dt}G_C = 0.0916(\Gamma_{P\Gamma R} - G_C) \quad (2.34)$$

2.2.5 Dalla Man model

Dalla Man's compartmental model is based on the principle of conservation of matter, from which a set of differential equations is obtained [10]. In general, the model is made up of six compartments (gastrointestinal tract, liver, glucose system, muscle and adipose tissue, beta cells, and insulin system). To represent patients with DM1, the compartment corresponding to beta cells is removed. The set of equations of Dalla Man model is shown in this section. In addition, for greater clarity, the equations are grouped into subsystems to which they belong. In Appendix A.1 the set of nominal values of the Dalla Man parameters are included.

Model equations

Glucose subsystem

$$\begin{cases} \dot{G}_p(t) = EGP(t) + Ra_{meal}(t) - U_{ii}(t) - E(t) - k_1 \cdot G_p(t) + k_2 \cdot G_t(t) \\ G_p(0) = G_{pb} \\ \dot{G}_t(t) = -U_{id}(t) + k_1 \cdot G_p(t) - k_2 \cdot G_t(t) \\ G_t(0) = G_{tb} \\ G(t) = G_p(t)/V_G \\ G(0) = G_b \end{cases} \quad (2.35)$$

Insulin subsystem

$$\begin{cases} \dot{I}_p(t) = -(m_2 + m_4) \cdot I_p(t) + m_1 \cdot I_l(t) + Ra_I(t) \\ I_p(0) = I_{pb} \\ \dot{I}_l(t) = -(m_1 + m_3) \cdot I_l(t) + m_2 \cdot I_p(t) \\ I_l(0) = I_{lb} \\ I(t) = I_p(t)/V_I \\ I(0) = I_b \end{cases} \quad (2.36)$$

Glucose rate of appearance

$$
\begin{cases}
Q_{sto}(t) = Q_{sto1}(t) + Q_{sto2}(t) \\
Q_{sto}(0) = 0 \\
\dot{Q}_{sto1}(t) = -k_{max} \cdot Q_{sto1}(t) + Dose \cdot \delta(t) \\
Q_{sto1}(0) = 0 \\
\dot{Q}_{sto2}(t) = -k_{empt}(Q_{sto}) \cdot Q_{sto2}(t) - k_{max} \cdot Q_{sto1}(t) \\
Q_{sto2}(0) = 0 \\
\dot{Q}_{gut}(t) = -k_{abs} \cdot (Q_{gut})(t) - k_{empt}(Q_{sto}) \cdot Q_{sto2}(t) \\
Q_{gut}(0) = 0 \\
Ra_{meal}(t) = \frac{f \cdot k_{abs} \cdot Q_{gut}(t)}{BW} \\
Ra_{meal}(0) = 0
\end{cases}
\tag{2.37}
$$

with

$$
k_{empt}(Q_{sto}) = k_{min} + \frac{(k_{max} - k_{min})}{2} \cdot \left\{ tanh\left[\alpha \left(Q_{sto} - \beta \cdot Dose \right) \right] \right.
$$
$$
\left. - tanh\left[\beta \left(Q_{sto} - c \cdot Dose \right) \right] + 2 \right\}
\tag{2.38}
$$

Endogenous glucose production

$$
EGP(t) = k_{p1}(t) - k_{p2} \cdot G_p(t)
$$
$$
- k_{p3}(t) \cdot X^L(t) + \xi \cdot X^H(t)
\tag{2.39}
$$

$$
\dot{X}^L(t) = -k_i \cdot \left[X^L(t) - I'(t) \right] X^L(0) = I_b
\tag{2.40}
$$

$$
\dot{I}'(t) = -k_i \cdot \left[I'(t) - I(t) \right] I'(0) = I_b
\tag{2.41}
$$

$$
\dot{X}^H(t) = -k_H \cdot X^H(t) + k_H \cdot max\left[\left(H(t) - H_b \right), 0 \right] X^H(0) = 0
\tag{2.42}
$$

Glucose utilization

$$
U_{ii}(t) = F_{cns}
\tag{2.43}
$$

$$
U_{id}(t) = \frac{k_{ir}(t) \cdot \left[V_{m0} + V_{mx}(t) \cdot X(t) \cdot \left(1 + r_1 \cdot risk \right) \right] \cdot G(t)}{K_{m0} + G_t(t)}
\tag{2.44}
$$

with

$$
\dot{X}(t) = -p_{2U} \cdot X(t) + p_{2U} \cdot \left[I(t) - I_b \right] \quad X(0) = 0
\tag{2.45}
$$

$$
risk = \begin{cases}
0 & if \ \ G \geq G_b \\
10 \cdot \left[f(G) \right]^2 & if \ \ G_{th} \leq G < G_b \\
10 \cdot \left[f(G_{th}) \right]^2 & if \ \ G < G_{th}
\end{cases}
\tag{2.46}
$$

$$
f(G) = \left[log(G) \right]^{r_2} - \left[log(G_b) \right]^{r_2}
\tag{2.47}
$$

Subcutaneous insulin kinetics

$$Ra_{Isc}(t) = k_{a1} \cdot I_{sc1}(t) + k_{a2} \cdot I_{sc1}(t) \tag{2.48}$$

with

$$\begin{cases} \dot{I}_{sc1}(t) = -(k_d + k_{a1}) \cdot I_{sc1}(t) + u_{sc}(t - \tau) \\ I_{sc1}(0) = I_{sc1ss} \\ \dot{I}_{sc2}(t) = k_d \cdot I_{sc1}(t) - k_{a2} \cdot I_{sc2}(t) \\ I_{sc2}(0) = I_{sc2ss} \end{cases} \tag{2.49}$$

Glucagon kinetics and secretion

$$\dot{H}(t) = -n \cdot H(t) + SR_H(t) + Ra_H(t) \qquad H(0) = H_b \tag{2.50}$$

with

$$SR_H(t) = SR_H^s(t) + SR_H^d(t) \tag{2.51}$$

$$\dot{SR}_H^s(t) = \begin{cases} -\rho \cdot \left[SR_H^s(t) - SR_H^b \right] & if \ \ G(t) \geq G_b \\ -\rho \cdot \left[SR_H^s(t) - max \left(\dfrac{\sigma \cdot \left[G_{th} - G(t) \right]}{I(t) + 1} + SR_H^b, 0 \right) \right] \\ & if \ \ G(t) < G_b \end{cases} \tag{2.52}$$

$$SR_H^d(t) = \delta \cdot max \left(-\frac{dG(t)}{dt}, 0 \right) \tag{2.53}$$

Subcutaneous glucagon kinetics

$$\begin{cases} \dot{H}_{sc1}(t) = -(k_{h1} + k_{h2}) \cdot H_{id1}(t) \\ H_{sc1}(0) = H_{sc1b} \\ \dot{H}_{sc2}(t) = k_{h1} \cdot H_{sc1}(t) - k_{h3} \cdot H_{sc2}(t) \\ H_{sc2}(0) = H_{sc2b} \end{cases} \tag{2.54}$$

$$Ra_H(t) = k_{h3} \cdot H_{sc2}(t) \tag{2.55}$$

2.3 Serial data

If it is possible to represent a vector of real numbers (measurements) with one or more elements at each point in time, then the state of the process can be represented as a vector of real-valued functions.

$$X = [x_1, x_2, ..., x_n] \tag{2.56}$$

The values x_i with $i = 1, ..., n$ at any moment are amplitudes of that component at that instant and the units of measure are units of amplitude.

The success of time series analysis depends on the construction of mathematical models that generate time functions under observations. The models must be built in such a way that the parameters of the models can be identified or related to the characteristics of the physical phenomenon. Hence, the process to obtain information from the observed data about the model parameters also provides information about the underlying process.

For this book, we focus primarily on model identification, parametric estimation, and classification on time series data. Below, we show the data used in this work as well as a description of them.

2.3.1 Continuous glucose monitoring system data

The data collected were obtained from a person with DM1, without the presence of any other disease or complication associated with the DM1, in a follow-up for 3 consecutive days of their normal routine, without exercise events as is shown in Fig. 2.2. Blood glucose figures were obtained using an insulin pump in real time with a continuous glucose monitoring system (CGMS) with the help of Apple's Healthkit. In addition, the amount of insulin was recorded in bolus administered, as well as the amount of carbohydrates for each food consumption. The glucose sensor was connected to the patient subcutaneously, which provides a sample of interstitial blood glucose in a time sample of 5 min from the moment of installation to 72 h later. The data provided was stored on a computer and used to obtain the dynamic behavior of glucose–insulin and carbohydrates. The graph of said data showed a predominance of glucose figures in hyperglycemic states of 158.3 mg/dl (±38.84 of), presenting the maximum peak of 271.4 mg/dl and a period of hypoglycemia at the beginning of the prolonged fasting of the third day of follow-up of 61.9 mg/dl, characteristic of an uncontrolled DM, where the ADA (American Diabetes Association) proposes figures from 70 to 130 mg/dl preprandials and less than 180 mg/dl postprandial, as well as an HBA1c <7% (154.2 mg/dl). Although the average figures are very close to the goals proposed by the ADA, the standard deviation is very high between the figures, so it suggests little stability of them. The therapeutic plan followed was the joint administration of food and insulin, identifying non-stable curve patterns, sometimes requiring extra doses of rescue insulin to lower postprandial serum glucose. Poor control was also demonstrated with the administration of insulin time after food, as well as multiple intakes with variations in amount of carbohydrates, the amount of insulin supplied to meet the glycemic demand being insufficient. It should be noted that the most stable period of the glycemic state occurred on the third day in prolonged fasting, where after a slight rise due to the intake of carbohydrates and insulin administered, the curve declined at 20 min prolonging to the hypoglycemia of 3 mg/dl at 3 h postprandial, which later recovered at 40 min and remained at figures around 8 mg/dl. This remained up to 6 h later until the optimal levels of 5 mg/dl fell before the new intake. The above can be explained by the cortisol regulation system that activates gluconeogenesis and was not activated until the hypoglycemia with stat levels of 6–7 mg/dl and approaching the time of the next intake.

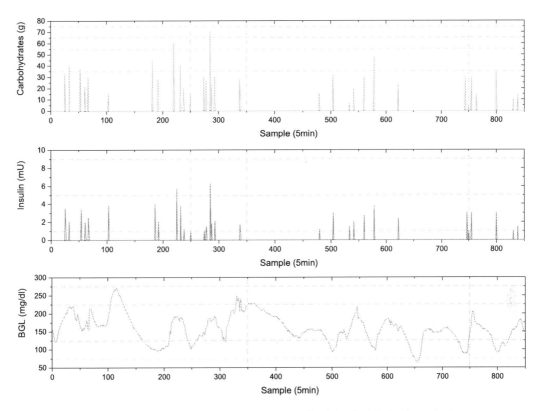

FIGURE 2.2 Patient historical data: carbohydrate consumption (yellow), insulin infusion (green), glucose concentration (red).

2.3.2 Tolerance tests to oral intake data

The normoglycemic glucose curve in Fig. 2.3 is characterized by a glycemia peak 30 min after loading, starting from a normal basal figure and reaching a maximum point of 155 mg/dl, which gradually decreases until reaching figures of less than 90 mg/dl at 2 h afterload.

The abnormal glycemic curve shown in Fig. 2.4 is characteristic of prediabetes, which shows postprandial hyperglycemia, starting with a normal basal glycemia and gradually increasing until reaching its maximum peak of 160 mg/dl at 2 h postload, maintaining and decreasing slowly over several minutes.

The abnormal glycemic curve is shown in Fig. 2.5, with glycemic dysregulation, which begins with a normal basal glycemia of 90 mg/dl that increases rapidly until it reaches its maximum peak of approximately 130 to 150 mg/dl during the first few minutes postload and drops equally rapidly within an hour afterload to approximate figures of 70 mg/dl. Subsequently, a slight increase is observed 2 h after loading around 100 mg/dl, which subsides and drops to normoglycemic levels. Despite the discrepancy in values, the metabolic factors that condition this situation could be explained by insufficient stimulation of in-

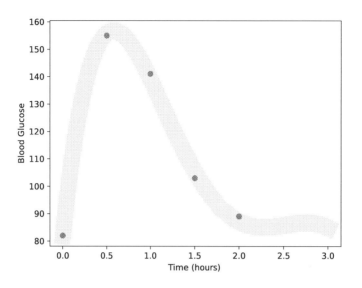

FIGURE 2.3 Data obtained from the glucose tolerance test of real patient a).

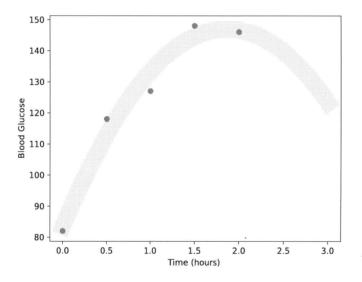

FIGURE 2.4 Data obtained from the glucose tolerance test of real patient b).

sulin secretion, which requires other endogenous stimuli, apart from carbohydrate intake, to maintain stable blood glucose levels, where gluconeogenesis could be activated to produce the second peak. Another possible explanation would be with extrinsic factors, where we would find the administration of fast-acting hypoglycemic agents or insulin analogs with or without new food intake, as well as considerable physical activity; all situations

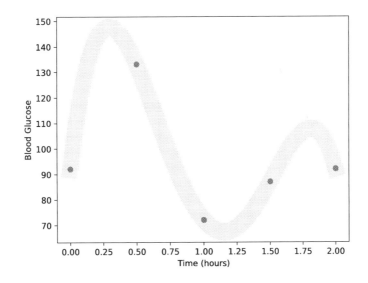

FIGURE 2.5 Data obtained from the glucose tolerance test of real patient c).

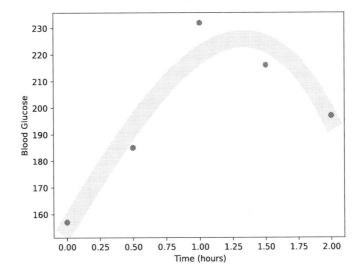

FIGURE 2.6 Data obtained from the glucose tolerance test of real patient d).

that are outside the controlled scope of the glucose tolerance test. Also, it should be noted that it is not possible to determine, using the ADA criteria, a diagnosis of diabetes or pre-diabetes despite the evident abnormality of the curve.

The abnormal glycemic curve shown in Fig. 2.6 is characteristic of uncontrolled diabetes, which shows sustained hyperglycemia, beginning with a high basal glycemia of 155 mg/dl that gradually increases until it reaches its maximum peak of 230 mg/dl in the first postload hour, and drops gradually without being able to reach the normoglycemic figures at 2 h post load (195 mg/dl).

References

[1] Bernard Tuch, Marjorie Dunlop, Joseph Proietto, Diabetes Research: A Guide for Postgraduates, CRC Press, 2003.

[2] Ravindranath Aathira, Vandana Jain, Advances in management of type 1 diabetes mellitus, World Journal of Diabetes 5 (5) (2014) 689.

[3] American Diabetes Association, et al., Standards of medical care in diabetes – 2017 abridged for primary care providers, Clinical Diabetes 35 (1) (2017) 5.

[4] M. Schönauer, A. Thomas, Sensor-augmented pump therapy – on the way to artificial pancreas, Avances en Diabetología 26 (3) (2010) 143–146.

[5] Ricardo Femat, Eduardo Ruiz-Velázquez, Griselda Quiroz, Weighting restriction for intravenous insulin delivery on T1DM patient via h_∞ control, IEEE Transactions on Automation Science and Engineering 6 (2) (2009) 239–247.

[6] Sara Trevitt, Sue Simpson, Annette Wood, Artificial pancreas device systems for the closed-loop control of type 1 diabetes: what systems are in development?, Journal of Diabetes Science and Technology 10 (3) (2016) 714–723.

[7] Eric Renard, Anne Farret, Jort Kropff, Daniela Bruttomesso, Mirko Messori, Jerome Place, Roberto Visentin, Roberta Calore, Chiara Toffanin, Federico Di Palma, et al., Day-and-night closed-loop glucose control in patients with type 1 diabetes under free-living conditions: results of a single-arm 1-month experience compared with a previously reported feasibility study of evening and night at home, Diabetes Care 39 (7) (2016) 1151–1160.

[8] G.P. Incremona, C. Cobelli, M. Messori, L. Magni, Individualized model predictive control for the artificial pancreas: In silico evaluation of closed-loop glucose control, IEEE Control Systems Magazine 38 (1) (2018) 86–104.

[9] John Thomas Sorensen, A physiologic model of glucose metabolism in man and its use to design and assess improved insulin therapies for diabetes, PhD thesis, Massachusetts Institute of Technology, 1985.

[10] Chiara Dalla Man, Francesco Micheletto, Dayu Lv, Marc Breton, Boris Kovatchev, Claudio Cobelli, The UVA/Padova type 1 diabetes simulator: new features, Journal of Diabetes Science and Technology 8 (1) (2014) 26–34.

[11] Troy Bremer, David A. Gough, Is blood glucose predictable from previous values? a solicitation for data, Diabetes 48 (3) (1999) 445–451.

[12] Meriyan Eren-Oruklu, Ali Cinar, Derrick K. Rollins, Lauretta Quinn, Adaptive system identification for estimating future glucose concentrations and hypoglycemia alarms, Automatica 48 (8) (2012) 1892–1897.

[13] Apurv Kamath, Aarthi Mahalingam, James Brauker, Methods of evaluating the utility of continuous glucose monitor alerts, 2010.

[14] Geoffrey McGarraugh, Alarm characterization for a continuous glucose monitor that replaces traditional blood glucose monitoring, 2010.

[15] Chiara Zecchin, Andrea Facchinetti, Giovanni Sparacino, Giuseppe De Nicolao, Claudio Cobelli, Neural network incorporating meal information improves accuracy of short-time prediction of glucose concentration, IEEE Transactions on Biomedical Engineering 59 (6) (2012) 1550–1560.

[16] Giovanni Sparacino, Francesca Zanderigo, Alberto Maran, Claudio Cobelli, Continuous glucose monitoring and hypo/hyperglycaemia prediction, Diabetes Research and Clinical Practice 74 (2006) S160–S163.

[17] Adiwinata Gani, Andrei V. Gribok, Srinivasan Rajaraman, W. Kenneth Ward, Jaques Reifman, Predicting subcutaneous glucose concentration in humans: data-driven glucose modeling, IEEE Transactions on Biomedical Engineering 56 (2) (2008) 246–254.

[18] Cesar C. Palerm, John P. Willis, James Desemone, B. Wayne Bequette, Hypoglycemia prediction and detection using optimal estimation, Diabetes Technology & Therapeutics 7 (1) (2005) 3–14.

[19] Daniel A. Finan, Cesar C. Palerm, Francis J. Doyle, Howard Zisser, Lois Jovanovic, Wendy C. Bevier, E. Dale Seborg, Identification of empirical dynamic models from type 1 diabetes subject data, 2008, pp. 2099–2104.

[20] Carmen Pérez-Gandía, A. Facchinetti, G. Sparacino, C. Cobelli, E.J. Gómez, M. Rigla, Alberto de Leiva, M.E. Hernando, Artificial neural network algorithm for online glucose prediction from continuous glucose monitoring, Diabetes Technology & Therapeutics 12 (1) (2010) 81–88.

[21] Scott M. Pappada, Brent D. Cameron, Paul M. Rosman, Raymond E. Bourey, Thomas J. Papadimos, William Olorunto, Marilyn J. Borst, Neural network-based real-time prediction of glucose in patients with insulin-dependent diabetes, Diabetes Technology & Therapeutics 13 (2) (2011) 135–141.

[22] B. Wayne Bequette, Continuous glucose monitoring: real-time algorithms for calibration, filtering, and alarms, Journal of Diabetes Science and Technology 4 (2) (2010) 404–418.

[23] Giovanni Sparacino, Andrea Facchinetti, Claudio Cobelli, "Smart" continuous glucose monitoring sensors: On-line signal processing issues, Sensors (Basel) 10 (2010) 6751.

[24] Peter J. Stuart, Shelley Crooks, Mark Porton, An interventional program for diagnostic testing in the emergency department, Medical Journal of Australia 177 (3) (2002) 131–134.

[25] Sebastian Pölsterl, Sailesh Conjeti, Nassir Navab, Amin Katouzian, Survival analysis for high-dimensional, heterogeneous medical data: Exploring feature extraction as an alternative to feature selection, Artificial Intelligence in Medicine 72 (2016) 1–11.

[26] Deepika Kumar, Nikita Jain, Aayush Khurana, Sweta Mittal, Suresh Chandra Satapathy, Roman Senkerik, Jude D. Hemanth, Automatic detection of white blood cancer from bone marrow microscopic images using convolutional neural networks, IEEE Access 8 (2020) 142521–142531.

[27] Ashir Javeed, Shijie Zhou, Liao Yongjian, Iqbal Qasim, Adeeb Noor, Redhwan Nour, An intelligent learning system based on random search algorithm and optimized random forest model for improved heart disease detection, IEEE Access 7 (2019) 180235–180243.

[28] N.N. Kulkarni, V.K. Bairagi, Extracting salient features for EEG-based diagnosis of Alzheimer's disease using support vector machine classifier, IETE Journal of Research 63 (1) (2017) 11–22.

[29] V. Anuja Kumari, R. Chitra, Classification of diabetes disease using support vector machine, International Journal of Engineering Research and Applications 3 (2) (2013) 1797–1801.

[30] Takanori Teramoto, Yuji Nonaka, Hiroki Tanaka, Satoshi Nakamura, Norihito Murayama, Tomoki Uchida, Takeshi Kanamori, Identifying glucose metabolism status in nondiabetic Japanese adults using machine learning model with simple questionnaire, Computational and Mathematical Methods in Medicine 2022 (2022) 1–10.

[31] Alexander Verner, Sumitra Mukherjee, An LSTM-based method for detection and classification of sensor anomalies, in: Proceedings of the 2020 5th International Conference on Machine Learning Technologies, 2020, pp. 39–45.

[32] Terrence J. Sejnowski, Charles R. Rosenberg, Parallel networks that learn to pronounce English text, Complex Systems 1 (1) (1987) 145–168.

[33] Wojciech Zaremba, Ilya Sutskever, Oriol Vinyals, Recurrent neural network regularization, arXiv preprint, arXiv:1409.2329, 2014.

[34] Shu Lih Oh, Yuki Hagiwara, U. Raghavendra, Rajamanickam Yuvaraj, N. Arunkumar, M. Murugappan, U. Rajendra Acharya, A deep learning approach for Parkinson's disease diagnosis from EEG signals, Neural Computing and Applications 32 (15) (2020) 10927–10933.

[35] G. Maragatham, Shobana Devi, LSTM model for prediction of heart failure in big data, Journal of Medical Systems 43 (5) (2019) 1–13.

[36] Richard N. Bergman, Y. Ziya Ider, Charles R. Bowden, Claudio Cobelli, Quantitative estimation of insulin sensitivity, American Journal of Physiology: Endocrinology and Metabolism 236 (6) (1979) E667.

[37] Roman Hovorka, Fariba Shojaee-Moradie, Paul V. Carroll, Ludovic J. Chassin, Ian J. Gowrie, Nicola C. Jackson, Romulus S. Tudor, A. Margot Umpleby, Richard H. Jones, Partitioning glucose distribution/transport, disposal, and endogenous production during IVGTT, American Journal of Physiology: Endocrinology and Metabolism 282 (5) (2002) E992–E1007.

[38] Morten Gram Pedersen, Gianna M. Toffolo, Claudio Cobelli, Cellular modeling: insight into oral minimal models of insulin secretion, American Journal of Physiology: Endocrinology and Metabolism 298 (3) (2010) E597–E601.

[39] Malgorzata E. Wilinska, Roman Hovorka, Simulation models for in silico testing of closed-loop glucose controllers in type 1 diabetes, Drug Discovery Today: Disease Models 5 (4) (2008) 289–298.

[40] Joseph Tiran, Kurt R. Galle, Daniel Porte, A simulation model of extracellular glucose distribution in the human body, Annals of Biomedical Engineering 3 (1975) 34–46.
[41] John R. Guyton, Richard O. Foster, J. Stuart Soeldner, Meng H. Tan, Charles B. Kahn, L. Koncz, Ray E. Gleason, A model of glucose-insulin homeostasis in man that incorporates the heterogeneous fast pool theory of pancreatic insulin release, Diabetes 27 (10) (1978) 1027–1042.
[42] Chiara Dalla Man, Robert A. Rizza, Claudio Cobelli, Meal simulation model of the glucose-insulin system, IEEE Transactions on Biomedical Engineering 54 (10) (2007) 1740–1749.
[43] Chis Oana-Teodora, Julio R. Banga, Eva Balsa-Canto, Structural identifiability of systems biology models: a critical comparison of methods, PLoS ONE 6 (11) (2011) e27755.
[44] Oana Chiş, Julio R. Banga, Eva Balsa-Canto, Genssi: a software toolbox for structural identifiability analysis of biological models, Bioinformatics 27 (18) (2011) 2610–2611.
[45] Tom Van Herpe, Marcelo Espinoza, Bert Pluymers, Ivan Goethals, Patrick Wouters, Greet Van den Berghe, Bart De Moor, An adaptive input–output modeling approach for predicting the glycemia of critically ill patients, Physiological Measurement 27 (11) (2006) 1057.
[46] Meriyan Eren-Oruklu, Ali Cinar, Lauretta Quinn, Donald Smith, Adaptive control strategy for regulation of blood glucose levels in patients with type 1 diabetes, Journal of Process Control 19 (8) (2009) 1333–1346.
[47] A. Karim El-Jabali, Neural network modeling and control of type 1 diabetes mellitus, Bioprocess and Biosystems Engineering 27 (2) (2005) 75–79.
[48] Alma Y. Alanis, Edgar N. Sanchez, Alexander G. Loukianov, Discrete-time adaptive backstepping nonlinear control via high-order neural networks, IEEE Transactions on Neural Networks 18 (4) (2007) 1185–1195.
[49] Oscar D. Sanchez, Alma Y. Alanis, E. Ruiz Velázquez, Roberto Valencia Murillo, Neural identification of type 1 diabetes mellitus for care and forecasting of risk events, Expert Systems with Applications (2021) 115367.
[50] Konstantia Zarkogianni, Konstantinos Mitsis, Eleni Litsa, M-T. Arredondo, G. Fico, Alessio Fioravanti, Konstantina S. Nikita, Comparative assessment of glucose prediction models for patients with type 1 diabetes mellitus applying sensors for glucose and physical activity monitoring, Medical & Biological Engineering & Computing 53 (12) (2015) 1333–1343.
[51] Ali Hassan El-Baz, Aboul Ella Hassanien, Gerald Schaefer, Identification of diabetes disease using committees of neural network-based classifiers, 2016, pp. 65–74.
[52] Marcos A. González-Olvera, Ana G. Gallardo-Hernández, Yu Tang, Maria Cristina Revilla-Monsalve, Sergio Islas-Andrade, A discrete-time recurrent neurofuzzy network for black-box modeling of insulin dynamics in diabetic type-1 patients, International Journal of Neural Systems 20 (02) (2010) 149–158.
[53] Stavroula Mougiakakou, Aikaterini Prountzou, Dimitra Iliopoulou, Konstantina Nikita, Andriani Vazeou, Christos Bartsocas, NN based glucose-insulin metabolism models for children with type 1 diabetes, in: Conference Proceedings: Annual International Conference of the IEEE Engineering in Medicine and Biology Society, vol. 1, IEEE Engineering in Medicine and Biology Society, 2006, pp. 3545–3548.
[54] W.A. Sandham, D.J. Hamilton, A. Japp, K. Patterson, Neural network and neuro-fuzzy systems for improving diabetes therapy 3 (1998) 1438–1441.
[55] Alessandro Aliberti, Irene Pupillo, Stefano Terna, Enrico Macii, Santa Di Cataldo, Edoardo Patti, Andrea Acquaviva, A multi-patient data-driven approach to blood glucose prediction, IEEE Access 7 (2019) 69311–69325.
[56] Chiara Zecchin, Andrea Facchinetti, Giovanni Sparacino, Giuseppe De Nicolao, Claudio Cobelli, A new neural network approach for short-term glucose prediction using continuous glucose monitoring time-series and meal information, in: 2011 Annual International Conference of the IEEE Engineering in Medicine and Biology Society, IEEE, 2011, pp. 5653–5656.
[57] Ananth Bhimireddy, Priyanshu Sinha, Bolu Oluwalade, Judy Wawira Gichoya, Saptarshi Purkayastha, Blood glucose level prediction as time-series modeling using sequence-to-sequence neural networks, in: CEUR Workshop Proceedings, 2020.
[58] Dinial Utami Nurul Qomariah, Handayani Tjandrasa, Chastine Fatichah, Classification of diabetic retinopathy and normal retinal images using CNN and SVM, in: 2019 12th International Conference on Information & Communication Technology and System (ICTS), IEEE, 2019, pp. 152–157.

[59] Eleonora Maria Aiello, Chiara Toffanin, Mirko Messori, Claudio Cobelli, Lalo Magni, Postprandial glucose regulation via KNN meal classification in type 1 diabetes, IEEE Control Systems Letters 3 (2) (2018) 230–235.

[60] Ratna Aminah, Adhi Harmoko Saputro, Diabetes prediction system based on iridology using machine learning, in: 2019 6th International Conference on Information Technology, Computer and Electrical Engineering (ICITACEE), IEEE, 2019, pp. 1–6.

[61] Ali Mohebbi, Tinna B. Aradottir, Alexander R. Johansen, Henrik Bengtsson, Marco Fraccaro, Morten Mørup, A deep learning approach to adherence detection for type 2 diabetics, in: 2017 39th Annual International Conference of the IEEE Engineering in Medicine and Biology Society (EMBC), IEEE, 2017, pp. 2896–2899.

[62] Arfan Ghani, Chan H. See, Vaisakh Sudhakaran, Jahanzeb Ahmad, Raed Abd-Alhameed, Accelerating retinal fundus image classification using artificial neural networks (ANNs) and reconfigurable hardware (FPGA), Electronics 8 (12) (2019) 1522.

[63] Laman R. Sultan, Diagnosis of type II diabetes based on feed forward neural network techniques, International Journal of Research in Pharmaceutical Sciences 11 (1) (2020) 1109–1116.

[64] Arkadiusz Kwasigroch, Bartlomiej Jarzembinski, Michal Grochowski, Deep CNN based decision support system for detection and assessing the stage of diabetic retinopathy, in: 2018 International Interdisciplinary PhD Workshop (IIPhDW), IEEE, 2018, pp. 111–116.

[65] Mamta Arora, Mrinal Pandey, Deep neural network for diabetic retinopathy detection, in: 2019 International Conference on Machine Learning, Big Data, Cloud and Parallel Computing (COMITCon), IEEE, 2019, pp. 189–193.

[66] Avula Benzamin, Chandan Chakraborty, Detection of hard exudates in retinal fundus images using deep learning, in: 2018 Joint 7th International Conference on Informatics, Electronics & Vision (ICIEV) and 2018 2nd International Conference on Imaging, Vision & Pattern Recognition (icIVPR), IEEE, 2018, pp. 465–469.

[67] S. Karthikeyan, Kumar P. Sanjay, R.J. Madhusudan, S.K. Sundaramoorthy, P.K. Krishnan Namboori, Detection of multi-class retinal diseases using artificial intelligence: an expeditious learning using deep CNN with minimal data, Biomedical & Pharmacology Journal 12 (3) (2019) 1577.

[68] Guanmin Chen, Peter Faris, Brenda Hemmelgarn, Robin L. Walker, Hude Quan, Measuring agreement of administrative data with chart data using prevalence unadjusted and adjusted kappa, BMC Medical Research Methodology 9 (2009) 1–8.

[69] Zolanda Anggraeni, Helmie Arif Wibawa, Detection of the emergence of exudate on the image of retina using extreme learning machine method, in: 2019 3rd International Conference on Informatics and Computational Sciences (ICICoS), IEEE, 2019, pp. 1–6.

[70] Tahira Nazir, Aun Irtaza, Zain Shabbir, Ali Javed, Usman Akram, Muhammad Tariq Mahmood, Diabetic retinopathy detection through novel tetragonal local octa patterns and extreme learning machines, Artificial Intelligence in Medicine 99 (2019) 101695.

[71] Kai He, Shuai Huang, Xiaoning Qian, Early detection and risk assessment for chronic disease with irregular longitudinal data analysis, Journal of Biomedical Informatics 96 (2019) 103231.

[72] M. Shanthi, Ramalatha Marimuthu, S.N. Shivapriya, R. Navaneethakrishnan, Diagnosis of diabetes using an extreme learning machine algorithm based model, in: 2019 IEEE 10th International Conference on Awareness Science and Technology (iCAST), IEEE, 2019, pp. 1–5.

Mathematical preliminaries

3.1 Evolutionary algorithms

3.1.1 Evonorm

The Evonorm evolutionary algorithm proposed in [1] replaced the crossing and mutation mechanisms with a distribution function. The Evonorm algorithm is described below:

Algorithm 1 Evonorm algorithm to solve minimization problems. f is the objective function, I_p the total number of particles and D_r the dimension of the problem.

1: $I_p, I_s \leftarrow$ define parameters

2: $P^i \leftarrow$ initialize i $\in I_p$ state of the particles randomly

3: **do**

4:      Choose particle with best swarm position $I_{x_{pr}}$

5:      Selection of the best individuals I_s from P

6:      **for** $i = 1 \rightarrow I_p$ **do**

7:          **for** $pr = 1 \rightarrow D_r$ **do**

8:             $\mu_{pr} = \frac{1}{I_s} \sum_{k=1}^{I_s} P_{S_{k,pr}}$

9:             $\sigma_{pr} = \sqrt{\frac{1}{I_s} \sum_{k=1}^{I_s} (P_{S_{k,pr}} - \mu_{pr})}$

10:

11:             $P_{i,pr} = \begin{cases} N(\mu_{pr}, \sigma_{pr}), & U(\bullet) > 0.5 \\ N(I_x, \sigma_{pr}), & otherwise \end{cases}$

12: **while** total number of iterations G is fulfilled

Set up a matrix P of size $I_p \times D_r$. Here, P is the population, I_p is the total number of individuals and D_r is the dimension. Each individual represents a solution. Then, the best I_s individuals from P are selected. The selected individual is a matrix P_S of size $I_s \times D_r$, where $I_s < I_p$, usually 10–20% of the total population. To generate a new population, the mean μ_{pr} and the standard deviation σ_{pr} are calculated using the best I_s individuals.

$$\mu_{pr} = \frac{1}{I_s} \sum_{k=1}^{I_s} P_{S_{k,pr}} \tag{3.1}$$

$$\sigma_{pr} = \sqrt{\frac{1}{I_s} \sum_{k=1}^{I_s} (P_{S_{k,pr}} - \mu_{pr})} \tag{3.2}$$

where the mean is μ_{pr} and the standard deviation is σ_{pr}, $pr \in \{1, 2, 3, ..., D_r\}$. To maintain a balance between exploration and exploitation, a stochastic process combined with heuristics is used to find new solutions that are not necessarily close to the mean. The individuals for the next generation are obtained using the best individual up to the moment I_x to intervene 50% in another case the average μ_{pr} is used:

$$P_{i,pr} = \begin{cases} N(\mu_{pr}, \sigma_{pr}), & U(\bullet) > 0.5 \\ N(I_{x_{pr}}, \sigma_{pr}), & otherwise \end{cases} \tag{3.3}$$

where $U(\bullet)$ is a uniform distribution function and $N(\bullet)$ is a random variable with a normal distribution.

3.1.2 Differential evolution

One of the most efficient algorithms for global optimization is the Differential Evolution (DE) algorithm, in general, it uses three operators to search the solution space: crossover, mutation, and selection [2].

The algorithm generates a population of N individuals (candidate solutions) randomly in the search space $S \subseteq \mathbb{R}^D$. The search behavior is established by the change of address and the difference step.

$$x_{i,g} = [x_{1,i,g}, x_{2,i,g}, ..., x_{D,i,g}] \in i = 1, 2, ..., N \tag{3.4}$$

For each generation g, individuals undergo mutation and crossover to generate a new population. Then, in the selection step, individuals in both populations compete to move onto the next generation.

For each individual $x_{i,g}$, a mutant vector $v_{i,g+1} = [v_{1,i,g}, ..., v_{D,i,g}]$ is generated, as follows:

$$v_{i,g} = x_{r1,g} + F(x_{r2,g} - x_{r3,g}) \tag{3.5}$$

where $r1, r2, r3 \in 1, 2, ..., N$ are numbers randomly chosen and are different from each other. $F \in [0, 1]$ is the crossover factor that controls the amplification of the differential expansion $x_{r2,g} - x_{r3,g}$ and N is at least four for the mutation to be applied. $x_{r1,g}$, the vector basis.

In the crossover stage, a test vector $u_{i,g+1} = [u_{1,i,g}, ..., u_{D,i,g}]$ is generated to increase diversification:

$$u_{j,i,g} = \begin{cases} v_{j,i,g}, & if \ rand \ [0, 1] \le CR \ or \ j = randn \\ x_{j,i,g}, & otherwise \ j = 1, 2, ..., D \end{cases} \tag{3.6}$$

where $CR \in [0, 1]$ is the crossover rate, $rand$ is a number uniformly distributed in the range $[0, 1]$, $randn$ is the randomly chosen index from the set $1, 2, ..., D$.

In the selection stage it is decided if the test vector $u_{i,g+1}$ passes to the next generation $(g + 1)$ by comparing the initial target individual $x_{i,g}$ with the following selection criteria:

$$x_{i,g+1} = \begin{cases} u_{i,g}, & if \ f(u_{i,g}) \le f(x_{i,g}) \\ x_{i,g}, & otherwise \end{cases} \tag{3.7}$$

where f is the objective function, $x_{i,g+1}$ is the individual in the new population [2].

Algorithm 2 DE algorithm to solve minimization problems where f is the objective function and $r1, r2, r3, ra$ are random numbers.

1: $F, C_R \leftarrow$ define parameters
2: $x_i \leftarrow$ initialize $i \in \{1, N\}$ individuals randomly such that $x_i \in \mathbb{R}^D$
3: **do**
4:     **for** $i = 1 \rightarrow N$ **do**
5:         $v_{i,g} \leftarrow x_{r_{1,g}} + F(x_{r_{2,g}} - x_{r_{3,g}})$ such that $r_{1,g} \neq r_{2,g} \neq r_{3,g} \neq i$
6:         **for** $j = 1 \rightarrow D$ **do**
7:             **if** $r_a \leq C_R$ **then**
8:                 $u_{i,j,g} \leftarrow v_{i,j,g}$
9:             **else**
10:                 $u_{i,j,g} \leftarrow x_{i,j,g}$
11:         **if** $f(u_{i,g}) < j(x_{i,g})$ **then**
12:             $x_{i,g} \leftarrow u_{i,g}$
13: **while** total number of generations G is fulfilled

3.2 Particle swarm-based algorithms

3.2.1 Particle swarm optimization

The Particle Swarm Optimization (PSO) algorithm is based on stochastic optimization techniques proposed in [3] and [4]. Its fundamental concept is based on the social interaction behavior of flocks and fish schooling. In general, PSO initializes a population randomly placed in the D-dimensional search space. The particles have two vectors, a velocity vector and a position vector, and are updated through these two vectors, learning from their best historical position and from the best global position achieved by the swarm in each generation. Let v_i and x_i be the velocity vector and position vector, respectively, with $i = 1, ..., N$, where N is the total number of particles in a population. The PSO algorithm is given by the equations:

$$v_i = w v_i + c_1 rand_1(x_i^b - x_i) + c_2 rand_2(x_i^g - x_i^j) \tag{3.8}$$

$$x_i = x_i + v_i \tag{3.9}$$

where $pBest_i$ are the best positions of the particle $i = (1, 2, ..., N)$, $gBest$ is the best position of the swarm, w is the inertia factor, c_1 and c_2 are parameters to weight the relative importance of x_i^b and x_i^g; respectively, $rand_1$ and $rand_2$ are randomly generated numbers uniformly distributed in $[0, 1]$.

PSO is one of the most popular optimization algorithms due to its ease and efficiency, which is why it has been used in various areas, *e.g.*, [5], [6], [7], [8].

Algorithm 3 PSO algorithm to solve minimization problems. f is the objective function, N the total number of particles and D the dimension of the problem.

1: W, $c_1, c_2 \leftarrow$ define parameters
2: $x_i \leftarrow$ initialize $i \in \{1, N\}$ particles randomly such that $x_i \in \mathbb{R}^D$
3: $v_i \leftarrow$ initialize $i \in \{1, N\}$ speeds randomly such that $v_i \in \mathbb{R}^D$
4: $x_i^b \leftarrow x_i$, initialization of best particle positions
5: **do**
6:     **for** $i = 1 \rightarrow N$ **do**
7:         **if** $f(x_i) < f(x_i^b)$ **then**
8:             $x_i^b \leftarrow x_i$
9:     Choose particle with best swarm position x^g
10:     **for** $i = 1 \rightarrow N$ **do**
11:         $v_i \leftarrow$ w $v_i + r_1 \, c_1(x_i^b - x_i) + r_2 \, c_2(x^g - x_i)$
12:         $x_i \leftarrow x_i + v_i$
13: **while** total number of iterations G is fulfilled

3.2.2 Ant colony optimization

The Ant Colony Optimization (ACO) algorithm, also called ANT, is inspired by the behavior of ants applied to solve optimization problems. The cooperation and communication is carried out through the generation of artificial pheromone trails [9]. The density of pheromones in real ants decreases due to evaporation, in ANT this effect is simulated with the application of the pheromone evaporation rule that is useful in artificial ant colonies.

In ANT, each ant is placed differently or in the same corners when generating the population. Eq. (3.10) (the probability equation) determines in which node each ant will be located at time (t):

$$P_{ij}^k(t) \begin{cases} \frac{[\tau_{ij}(t)]^\alpha [\eta_{ij}(t)]^\beta}{\sum_{i \in N_i} [\tau_{ij}(t)]^\alpha \eta_{ij}(t)]^\beta}, & for \ k \\ 0, & otherwise \end{cases} \tag{3.10}$$

where $\tau_{ij}(t)$ is the amount of pheromones traced in the corners (i, j). η_{ij} is the visibility value between corners (i, j). α is the relevance of the pheromone trail to the problem. β is the importance to the visibility value. N_i represents the set of node points that have not been chosen.

The ants make choices according to the probability equation. The iteration is completed when all problem nodes have been visited, then the pheromone trace is updated according to the following equation [10]:

$$\tau_{ij}(t + n) = (1 - \rho)\tau_{ij}(t) + \Delta\tau_{ij}(t) \tag{3.11}$$

where ρ is the coefficient of evaporation $(0 < \rho < 1)$. τ_{ij} is the amount of pheromones due to the path of the ants, it is calculated by the following equation:

$$\Delta\tau_{ij}^{k}(t) = \sum_{k=1}^{m} \tau_{ij} \tag{3.12}$$

where m is the number of ants. τ_{ij}^{k} is the amount of pheromone left by ant k at the (i, j) corner.

The amount of pheromones contributed by the ant (k) is represented by the following equation [11]:

$$\Delta\tau_{ij}^{k} = \frac{Q}{L_k} \tag{3.13}$$

where Q is a constant and L_k is the length of the ant's path k.

3.3 Neural networks

This section shows the neural networks used in this work. Simple architectures include the multilayer perceptron neural network and the high-order recurrent neural network. Below, is a brief description of these neural networks.

3.3.1 Multilayer perceptron neural network

The Multilayer Perceptron (MLP) is one of the most influential neural network models used in various problems due to its efficiency and flexibility [12,13], such as, for example, in classification applications [13]. It consists of an input layer, followed by one or more hidden layers, and an output layer of interconnected nodes. Dense layers are composed of two or more hidden layers with their nodes connected to other nodes of neighboring hidden layers by means of weights. In general, MLP is a feedforward neural network, which means that information flows in one direction, from the input layer through the hidden layers to the output layer.

The following equation describes the output of the neuron j of the hidden layers:

$$s_j = \sigma\left(\sum_{i=1}^{n} w_{ji}x_i + b_i\right) \tag{3.14}$$

where σ is the activation function, b_i are layers of the layer, and w_{ji} are weights.

Then, the output of the *and* network is obtained by the following equation:

$$y = \sigma\left(\sum_{j=1}^{m} w_{kj}s_i + b_0\right) \tag{3.15}$$

where σ is the activation function, w_{kj} are pesos, and b_0 are biases.

Algorithm 4 ANT algorithm to solve minimization problems. N the total number of particles and n the dimension of the problem.

1: $n =$ number of dimensions
2: Divide the i-th dimension in to $B_i - 1$ intervals, $i \in [1, n]$
3: $\alpha =$ importance of pheromone amounts
4: $Q =$ deposition constant
5: $\rho =$ evaporation rate $\in (0, 1)$
6: $\tau_{i, j_i} = \tau_0$ (initial pheromone) for $i \in [1, n]$ and $j_i \in [1, B_i - 1]$
7: Randomly initialize a population of ants (candidate solutions) $a_k, k \in [1, N]$
8: **while** not (termination criterion) **do**
9:      **for** each ant $a_k, k \in [1, N]$ **do**
10:         **for** each dimension $i \in [1, n]$ **do**
11:             **for** each discretized interval $[b_{ij}, b_{i,j+1}], j \in [1, B_i - 1]$ **do**
12:                 Probability $p_{ij}^{(k)} \leftarrow \tau_{ij}^\alpha / \sum_{m=1}^{B_i - 1} \tau_{im}^\alpha$
13:             Next discretized interval
14:             $a_k(x_i) \leftarrow U[b_{ij}, b_{i,j+1}]$ with probability $p_{ij}^{(k)}$
15:         Next dimension
16:     Next ant
17:     $L_k \leftarrow$ cost of solution constructed by ant $a_k, k \in [1, N]$
18:     **for** each dimension $i \in [1, n]$ **do**
19:         **for** each discretized interval $[b_{ij}, b_{i,j+1}], j \in [1, B_i - 1]$ **do**
20:             **for** each ant $a_k, k \in [1, N]$ **do**
21:                 **if** $a_k(x_i) \in [b_{ij}, b_{i,j+1}]$ **then**
22:                     $\Delta \tau_{ij}^{(k)} \leftarrow Q/L_k$
23:                 **else**
24:                     $\Delta \tau_{ij}^{(k)} \leftarrow 0$
25:             Next ant
26:             $\tau_{ij} \leftarrow (1 - \rho)\tau_{ij} + \sum_{k=1}^{N} \Delta \tau_{ij}^{(k)}$
27:         Next discretized interval
28:     Next dimension
29: Next generation

3.3.2 Discrete-time high order neural networks

Due to the nonlinear characteristics, neural networks can solve modeling problems. In fact, interest in the use of recurrent neural networks for identification and modeling has grown recently [14]. Recurrent structures facilitate modeling by focusing simply on the selection of inputs to the network, the number of hidden units, type of activation function, and training algorithm.

One of the useful recurrent neural networks in control is the high-order recurrent neural network (RHONN) [15]. In particular, the neural network of high order in discrete time, for which it is considered the obtaining of an affine nonlinear system model of the form:

$$x(k + 1) = F(x(k)) + G(x(k))u(k) \tag{3.16}$$

$$y(k) = H(x(k)) \tag{3.17}$$

where $x \in \mathbb{R}^n$, $u \in \mathbb{R}^m$, $F \in \mathbb{R}^n \to \mathbb{R}^n$, and $G \in \mathbb{R}^n \to \mathbb{R}^{n \times m}$ are smooth mappings. Then, RHONN in discrete time can be represented by the following equation:

$$\hat{x}(k + 1) = w_i^T z_i(x(k), u(k)) \; i = 1, ..., n \tag{3.18}$$

where $\hat{x}(i = 1, 2, ..., n)$ is the state of the ith neuron, n is the state dimension, L_i is the number of higher-order connections, $\{I_1, I_2, ..., I_i\}$ is a collection of unordered subsets of $\{i = 1, 2, ..., n + m\}$, m is the number of entries, $w_i (i = 1, 2, ..., n)$ are the weights, and $z_i(x(k), u(k))$ is obtained by the following equation:

$$z_i(x(k), u(k)) = \begin{bmatrix} z_{i_1} \\ z_{i_2} \\ \vdots \\ z_{i_{L_i}} \end{bmatrix} = \begin{bmatrix} \prod_{j \in I_1} \xi_{ij}^{d_{ij}^{(1)}} \\ \prod_{j \in I_2} \xi_{ij}^{d_{ij}^{(2)}} \\ \vdots \\ \prod_{j \in I_{L_i}} \xi_{ij}^{d_{ij}^{(L_i)}} \end{bmatrix} \tag{3.19}$$

with $d_{i_j}(k)$ being non-negative integers, ξ_i can be defined as:

$$\xi_i = \begin{bmatrix} \xi_{i_1} \\ \vdots \\ \xi_{i_1} \\ \vdots \\ \xi_{i_{n+1}} \\ z_{i_{n_x+m}} \end{bmatrix} = \begin{bmatrix} S(x_1) \\ \vdots \\ S(x_{x_n}) \\ u_1 \\ \vdots \\ u_m \end{bmatrix} \tag{3.20}$$

where $u = [u_1, u_2, .., u_m]$ is the input vector, and $S(\cdot)$ is:

$$S(\varsigma) = \mu_i tanh(\beta_i \varsigma) \tag{3.21}$$

where ς is a real valued variable and μ, β are positive constants.

3.3.2.1 *RHONN training algorithm*
The objective of the training is to find the optimal weight vector that minimizes the approximation error. Typically, RHONN is input with the Extended Kalman Filter (EKF) training

algorithm described in [15]. EKF for training converts the weights of the network into states to be estimated, in this way the error between the output of the neural network and the output of the measured plant is considered as white noise and because the mapping of the neural network is not linear. EKF is described below:

$$K_i(k) = P_i(k)H_i(k)M(k)$$
$$w_i(k+1) = w_i(k) + \eta_i K_i(k)e(k) \tag{3.22}$$
$$P_i(k+1) = P_i(k) - K_i(k)H_i^T P_i(k) + Q_i(k)$$

with

$$M(k) = [R_i(k) + H_i^T(k)P_i(k)H_i]^{-1} \tag{3.23}$$

$$e(k) = [y(k) - \hat{y}(k)] \tag{3.24}$$

where $P_i \in \Re^{L_i \times L_i}$ is the covariance matrix of the error, $w_i \in \Re^{L_i}$ is the weight vector, L_i is the number of weights in the neural network, $y \in \Re^m$ is the measured output vector, $\hat{y} \in \Re^m$ is the output of the network, η is a design parameter, $K_i \in \Re^{L_i \times m}$ is the Kalman gain matrix, $Q_i \in \Re^{L_i \times L_i}$ is the covariance matrix of the state noise, $R_i \in \Re^{m \times m}$ is the covariance matrix of the measurement noise, and $H_i \in \Re^{L_i \times L_i}$ is a matrix for which each entry (H_{ij}) is computed as follows:

$$H_{ij}(k) = \left[\frac{\partial \hat{y}(k)}{\partial w_{ij}(k)} \right]_{w_i(k) = \hat{w}_i(k+1)} \quad i = 1, ..., n \ j = 1, ..., L_i \tag{3.25}$$

3.4 Deep neural networks

Shallow architectures of neural networks limit the learning of complex nonlinear relationships [16]. Hence, to solve this problem, neural networks increase the number of hidden layers of nonlinear operations that allow capturing complex dynamics through deep training [17]. These networks are known as Deep Neural Networks (DNN).

Models generated by deep neural networks are a promising approach for automatic feature extraction from complex data model representations [18]. Furthermore, through the data obtained by sensors, the DNN can contribute to developing better and more reliable diagnosis and disease classification systems.

3.4.1 Convolutional neural network

One of the best known deep neural networks is called the Convolutional Neural Network (CNN), which has been used successfully in different areas such as object recognition, image classification, and recently in time series classification [19–21].

Generally, CNN has applications in information that can be represented in 2D matrices. However, recently, 1D architectures in the convolution layer have been used for classification. In 1D architecture, 1D kernels are operated in the convolution layer and 1D filters are used on the input signal.

The convolutional neural network is made up of three layers, a filter bank layer, a non-linearity layer, and a feature grouping layer [22].

The filter bank or convolution layer is composed of filters applied over the input layer. The output of this layer corresponds to the convolution of the weights of the neurons with the input. The result is a new feature map set [23].

In the nonlinearity layer, various nonlinear activation functions are applied to limit or cut off the output. Commonly, the CNN uses the Rectified Linear Unit (ReLU) activation function as shown below:

$$ReLU = \begin{cases} 0, & if \quad x < 0, \\ x, & if \quad x \geq 0 \end{cases} \tag{3.26}$$

The feature extraction layer is used to reduce the dimension of the data, the methods used by CNN are maximum pooling and mean pooling.

Finally, in the output layer, the softmax layer is used to increase the maximum probability of the output. The softmax function is presented below:

$$o_i = \frac{e^{z_i}}{\sum_{i=1}^{N} e^{z_i}} \tag{3.27}$$

where o is the output after the softmax layer, z is the output of the neural network, and N represents the number of classes or outputs.

3.4.2 Long short-term memory recurrent neural network

The Long Short-Term Memory Recurrent LSTM neural network was proposed in [24] to solve the vanishing gradient problem presented by recurrent neural networks. The LSTM introduces the forget gate to control the memory of past states. In general, the LSTM can be described by the following equations [24]:

$$i_t = \sigma(W_{xi}x_t + W_{hi}h_{t-1} + W_{ci}c_{t-1} + bi) \tag{3.28}$$

$$f_t = \sigma(W_{xf}x_t + W_{hf}h_{t-1} + W_{cf}c_{t-1} + bf) \tag{3.29}$$

$$\tilde{C}_t = tanh(W_{xc}x_t + W_{hc}h_{t-1} + bc) \tag{3.30}$$

$$C_t = f_t C_{t-1} + i_t \tilde{C}_t \tag{3.31}$$

$$o_t = \sigma(W_{xo}x_t + W_{ho}h_{t-1} + W_{co}C_t + bo) \tag{3.32}$$

$$h_t = o_t tanh(C_t) \tag{3.33}$$

where the Ws are weight matrices, bs are biases, σ is the sigmoid logistic function. f, i, o are the forget, input, and output gates, respectively. C is the cell activation, ht is a value scaled by the $tanh$ function between -1 and 1 of the output gate.

3.4.3 Bidirectional LSTM

Bidirectional Recurrent Neural Networks (BRNN) were first introduced in [25], with the aim of dividing the recurrent neural network state into a positive direction (forward state) and a negative direction (regress state). which are independent of each other [25]. This bidirectional structure is applied in the LSTM to access the state from both directions (past and future).

Then, the forward state $\overrightarrow{h_t}$ and the backward state $\overleftarrow{h_t}$ are combined to produce the output y_t. where the hidden states h is a LSTM cell. The set of BiLSTM equations is shown below:

$$\overrightarrow{h_t} = H(W_{x\overrightarrow{h}}x_t + W_{\overrightarrow{h}\overrightarrow{h}}\overrightarrow{h}_{t-1} + b_{\overrightarrow{h}}) \tag{3.34}$$

$$\overleftarrow{h_t} = H(W_{x\overleftarrow{h}}x_t + W_{\overleftarrow{h}\overleftarrow{h}}\overleftarrow{h}_{t-1} + b_{\overleftarrow{h}}) \tag{3.35}$$

$$y_t = W_{\overrightarrow{h}y}\overrightarrow{h}_t + W_{\overleftarrow{h}y}\overleftarrow{h}_t + b_y \tag{3.36}$$

The training of the bidirectional LSTM is carried out for the state of advance as for the state of retreat. There is no interaction between them during training so it can be trained with the same algorithms as a unidirectional LSTM. The bidirectional structure is shown in Fig. 3.1.

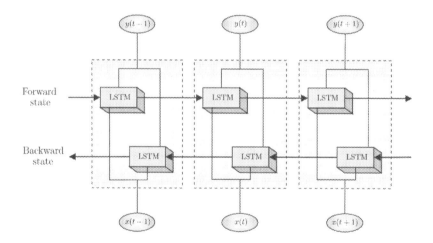

FIGURE 3.1 Bi-LSTM architecture.

3.4.4 LSTM fully convolutional networks

The LSTM-FCN deep neural network proposed in [26] is useful for classifying time series data, and is composed of two blocks, these are the convolutional block and the LSTM block. Both blocks receive the same time series as input. The LSTM-FCN is shown in Fig. 3.2.

There are several configurations in the convolution block, in general 128, 256, and 128 are used, respectively, as proposed in [27] and followed by the ReLU activation function. The global average pooling is used in the feature extraction layer.

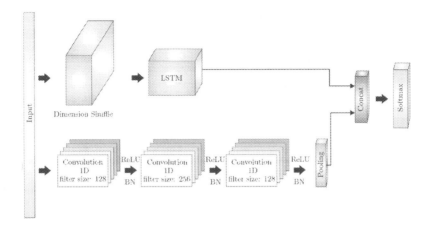

FIGURE 3.2 LSTM-FCN architecture.

For the LSTM block, the time series is dimensioned and passes through this block. Then, the output of the first and second blocks are concatenated to generate a multiclass classification through a softmax [28] layer.

3.4.5 Residual deep networks

The deep neural network Residual Network (ResNet) presented in [29], is composed of stacked "residual units". The general form of the ResNet is presented below:

$$y_l = x_l + F(x_l) \tag{3.37}$$
$$x_{l+1} = \theta(y_l)$$

where x_l is the input, $x_l + 1$ is the output of unit l, $\mathcal{F}$ represents the residual function. θ is the ReLU activation function [29]. $F(x_l) + x_l$ is achieved by feedforward neural networks with "shortcut connections", see Fig. 3.3.

ResNet is made up of more than 100 deep layers, thus it has demonstrated remarkable accuracy in complex recognition tasks [30,31]. The main concept of ResNet is to learn the additive residual function F relative to x_l. In this way, a shortcut or jump is achieved to move between layers via "hop connections".

References

[1] L. Torres, Evonorm, a new evolutionary algorithm to continuous optimization, in: Workshop on Optimization by Building and Using Probabilistic Models (OBUPM 2006), Genetic and Evolutionary Computation Conference (GECCO 2006), CD Proceeding Tutorials and Workshops, Seattle, 2006.

[2] Bo Peng, Bo Liu, Fu-Yi Zhang, Ling Wang, Differential evolution algorithm-based parameter estimation for chaotic systems, Chaos, Solitons and Fractals 39 (5) (2009) 2110–2118.

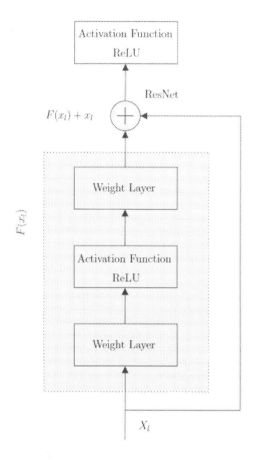

FIGURE 3.3 Building block of a ResNet.

[3] Russell Eberhart, James Kennedy, A new optimizer using particle swarm theory, in: MHS'95. Proceedings of the Sixth International Symposium on Micro Machine and Human Science, IEEE, 1995, pp. 39–43.

[4] James Kennedy, Russell Eberhart, Particle swarm optimization, in: Proceedings of ICNN'95 – International Conference on Neural Networks, vol. 4, IEEE, 1995, pp. 1942–1948.

[5] Ho Shinn-Ying, Lin Hung-Sui, Liauh Weei-Hurng, Ho Shinn-Jang Opso, Orthogonal particle swarm optimization and its application to task assignment problems, IEEE Transactions on Systems, Man and Cybernetics. Part A. Systems and Humans 38 (2) (2008) 288–298.

[6] Tridib Kumar Das, Ganesh Kumar Venayagamoorthy, Usman O. Aliyu, Bio-inspired algorithms for the design of multiple optimal power system stabilizers: SPPSO and BFA, IEEE Transactions on Industry Applications 44 (5) (2008) 1445–1457.

[7] Yamille Del Valle, Ganesh Kumar Venayagamoorthy, Salman Mohagheghi, Jean-Carlos Hernandez, Ronald G. Harley, Particle swarm optimization: basic concepts, variants and applications in power systems, IEEE Transactions on Evolutionary Computation 12 (2) (2008) 171–195.

[8] Mark P. Wachowiak, Renata Smolíková, Yufeng Zheng, Jacek M. Zurada, Adel Said Elmaghraby, An approach to multimodal biomedical image registration utilizing particle swarm optimization, IEEE Transactions on Evolutionary Computation 8 (3) (2004) 289–301.

[9] T.T. Ergüzel, E. Akbay, ACO (ant colony optimization) algoritması ile yörünge takibi, UMES, İzmit, Kocaeli, 2007.

[10] A. Colornmi, Distributed optimization by ant colonies, in: Proceedings of the First European Conference on Artificial Life, The MIT Press, 1992.

[11] Kemal Alaykiran, Orhan Engin, Karinca kolonileri metasezgiseli ve gezgin satici problemleri üzerinde bir uygulamasi, Gazi Üniversitesi Mühendislik Mimarlık Fakültesi Dergisi 20 (1) (2005).

[12] Xiaojun Zhai, Amine Ait Si Ali, Abbes Amira, Faycal Bensaali, MLP neural network based gas classification system on Zynq SoC, IEEE Access 4 (2016) 8138–8146.

[13] T.Y. Lim, M.M. Ratnam, M.A. Khalid, Automatic classification of weld defects using simulated data and an MLP neural network, Insight – Non-Destructive Testing and Condition Monitoring 49 (3) (2007) 154–159.

[14] Alma Y. Alanis, Blanca S. Leon, Edgar N. Sanchez, Eduardo Ruiz-Velazquez, Blood glucose level neural model for type 1 diabetes mellitus patients, International Journal of Neural Systems 21 (06) (2011) 491–504.

[15] Edgar N. Sanchez, Alma Y. Alanís, Alexander G. Loukianov, Discrete-Time High Order Neural Control, Springer, 2008.

[16] Feng Jia, Yaguo Lei, Jing Lin, Xin Zhou, Na Lu, Deep neural networks: A promising tool for fault characteristic mining and intelligent diagnosis of rotating machinery with massive data, Mechanical Systems and Signal Processing 72 (2016) 303–315.

[17] Olalekan Ogunmolu, Xuejun Gu, Steve Jiang, Nicholas Gans, Nonlinear systems identification using deep dynamic neural networks, arXiv preprint, arXiv:1610.01439, 2016.

[18] Hui Wen Loh, Chui Ping Ooi, Elizabeth Palmer, Prabal Datta Barua, Sengul Dogan, Turker Tuncer, Mehmet Baygin, U. Rajendra Acharya, GaborPDNet: Gabor transformation and deep neural network for Parkinson's disease detection using EEG signals, Electronics 10 (14) (2021) 1740.

[19] Ross Girshick, Jeff Donahue, Trevor Darrell, Jitendra Malik, Rich feature hierarchies for accurate object detection and semantic segmentation, in: Proceedings of the IEEE Conference on Computer Vision and Pattern Recognition, 2014, pp. 580–587.

[20] Yi Zheng, Qi Liu, Enhong Chen, Yong Ge, J. Leon Zhao, Time series classification using multi-channels deep convolutional neural networks, in: International Conference on Web-Age Information Management, Springer, 2014, pp. 298–310.

[21] Alex Krizhevsky, Ilya Sutskever, Geoffrey E. Hinton, ImageNet classification with deep convolutional neural networks, Advances in Neural Information Processing Systems 25 (2012) 1097–1105.

[22] Yann LeCun, Koray Kavukcuoglu, Clément Farabet, Convolutional networks and applications in vision, in: Proceedings of 2010 IEEE International Symposium on Circuits and Systems, IEEE, 2010, pp. 253–256.

[23] Saad Albawi, Oguz Bayat, Saad Al-Azawi, Osman N. Ucan, Social touch gesture recognition using convolutional neural network, Computational Intelligence and Neuroscience (2018) 2018.

[24] Sepp Hochreiter, Jürgen Schmidhuber, Long short-term memory, Neural Computation 9 (8) (1997) 1735–1780.

[25] Mike Schuster, Kuldip K. Paliwal, Bidirectional recurrent neural networks, IEEE Transactions on Signal Processing 45 (11) (1997) 2673–2681.

[26] B. Koley, Debangshu Dey, An ensemble system for automatic sleep stage classification using single channel EEG signal, Computers in Biology and Medicine 42 (12) (2012) 1186–1195.

[27] Fazle Karim, Somshubra Majumdar, Houshang Darabi, Shun Chen, LSTM fully convolutional networks for time series classification, IEEE Access 6 (2017) 1662–1669.

[28] Yunbin Kim, Jaewon Sa, Yongwha Chung, Daihee Park, Sungju Lee, Resource-efficient pet dog sound events classification using LSTM-FCN based on time-series data, Sensors 18 (11) (2018) 4019.

[29] Kaiming He, Xiangyu Zhang, Shaoqing Ren, Jian Sun, Deep residual learning for image recognition, in: Proceedings of the IEEE Conference on Computer Vision and Pattern Recognition, 2016, pp. 770–778.

[30] Olga Russakovsky, Jia Deng, Hao Su, Jonathan Krause, Sanjeev Satheesh, Sean Ma, Zhiheng Huang, Andrej Karpathy, Aditya Khosla, Michael Bernstein, et al., ImageNet large scale visual recognition challenge, International Journal of Computer Vision 115 (3) (2015) 211–252.

[31] Lin Tsung-Yi, Michael Maire, Serge Belongie, James Hays, Pietro Perona, Deva Ramanan, Piotr Dollár, C. Lawrence Zitnick, Microsoft coco: Common objects in context, in: European Conference on Computer Vision, Springer, 2014, pp. 740–755.

4

Parameter estimation for glucose–insulin dynamics

A critical issue for the use of physiological models is the parametric identification of a model for an individual or a population [1]. It is common that the parametric characteristics of the internal behavior of systems, above all biological and physiological, are not measurable [1]. Hence, these measurements are approximate through parametric estimation. These parameters can be estimated by characterizing the system through time series obtained by sensor measurements of one or more of the system output variables. Parametric estimation is not a simple task, mainly because biological or physiological models contain a large number of parameters and some of them cannot be estimated.

Parametric identification for DM1 models has been performed to represent different scenarios of this disease [2], in this way physiologically motivated time-varying models for glucose regulation are built. The most common models to carry out the task of parametric estimation of diabetes mellitus have been the models proposed by Hovorka, Dalla Man, Sorensen, and UVA/Padova [2,3].

In this book, the parametric adjustment of the Sorensen and Dalla Man models is presented in such a way that it coincides with the output of a continuous glucose monitoring sensor of a patient with DM1. Put another way, the parameters of the models are successively adjusted by an estimation problem, particularly using optimization algorithms until the error of the output of the models and the series of data obtained by a sensor is minimized.

Fig. 4.1 shows the identification process carried out by the heuristic algorithm. Where the unknown system is a patient with DM1, $u(t)$ is the input data and the output caused $y(t)$ according to the unknown inputs and disturbances. $\hat{y}(t)$ is the output of the Sorensen and Dalla Man mathematical models. $e(t)$ is the error obtained by $y(t) - \hat{y}(t)$.

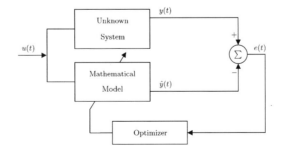

FIGURE 4.1 Representative diagram of identification with evolutionary algorithms.

Bio-Inspired Strategies for Modeling and Detection in Diabetes Mellitus Treatment
https://doi.org/10.1016/B978-0-44-322341-9.00013-6

4.1 Affine system

The Sorensen model and the Dalla Man model can be represented in the general form of the nonlinear system in the affine form:

$$\Sigma \begin{cases} \dot{x} = f(x, p) + g(x, p)u \\ y = h(x) \end{cases} \tag{4.1}$$

where, $x = \{x_1, \ldots, x_n\} \in \mathbb{R}^n$ is the state variable, $p = \{p_1, \ldots, p_q\} \in \mathbb{R}^q$ is the unknown/uncertain parameter vector in parameter space P, $u = \{u_1, \ldots, u_{n_u}\} \in \mathbb{R}^{n_u}$ is the input (control) vector, $y = \{y_1, \ldots, y_m\} \in \mathbb{R}^m$ is the m-dimensional output. $f(x, p)$ is a smooth nonlinear function. The state equation of the models can be written as:

4.1.1 Affine system of the Dalla Man model

To carry out the parametric estimation of the Dalla Man model, the variables can be represented in the state space by defining the following set of equations:

$$\dot{x}_1 = EGP + R_a - Uii - E - k_1 * x_1 + k_2 * x_2$$
$$\dot{x}_2 = -Uid + k_1 * x_1 - k_2 * x_2$$
$$\dot{x}_3 = -(m_1 + m_3) * x_3 + m_2 * x_4$$
$$\dot{x}_4 = -(m_2 + m_4) * x_4 + m_1 * x_3 + R_{ai}$$
$$\dot{x}_5 = -k_i(x_5 - \frac{x_4}{V_I})$$
$$\dot{x}_6 = k_i(x_6 - \frac{x_4}{V_I})$$
$$\dot{x}_7 = -k_{gri} * x_7 + D(t)\delta(t)$$
$$\dot{x}_8 = -k_{empt} * x_8 + k_{gri} * x_7$$
$$\dot{x}_9 = -k_{abs} * x_9 + k_{empt} * x_8 \tag{4.2}$$
$$\dot{x}_{10} = -p_{2U}(I_b - \frac{x_4}{V_I}) - p_{2U}x_{10}$$
$$\dot{x}_{11} = -\rho * (x_{11} - max((\sigma * (G_{th} - G)/(max(I - I_{th}, 0))) + SR_H^b, 0))$$
$$\dot{x}_{12} = -n * x_{12} + SR_H + Ra_h$$
$$\dot{x}_{13} = -k_H * x_{13} + k_H * max(x_{12} - Hb, 0)$$
$$\dot{x}_{14} = -(k_d + k_{a1}) * x_{14} + IIR$$
$$\dot{x}_{15} = k_d * x_{14} - k_{a2} * x_{15}$$
$$\dot{x}_{16} = -(k_{h1} + k_{h2}) * x_{16}$$
$$\dot{x}_{17} = k_{h1} * x_{16} - k_{h3} * x_{17}$$

where $x_1 = G_p$, $x_2 = G_t$, $x_3 = I_l$, $x_4 = I_p$, $x_5 = I_1$, $x_6 = I_d$, $x_7 = Q_{sto1}$, $x_8 = Q_{sto2}$, $x_9 = Q_{gut}$, $x_{10} = X$, $x_{11} = SR_H^s$, $x_{12} = H$, $x_{13} = X^H$, $x_{14} = I_{sc1}$, $x_{15} = I_{sc2}$, $x_{16} = H_{sc1}$ $x_{17} = H_{sc2}$ $y = G$, $u = IIR$.

4.1.2 Affine system of Sorensen model

To carry out the parametric estimation of the Sorensen model, the variables can be represented in the state space by defining the following set of equations:

$$\dot{x}_1 = p_1(x_3 - x_1) - p_2(x_1 - x_2)$$

$$\dot{x}_2 = p_3(x_1 - x_2) - p_4 p_{76}$$

$$\dot{x}_3 = p_5(p_6 x_1 + p_7 x_5 + p_8 x_6 + p_9 x_7 - p_{10}x_3 - p_{77})$$

$$\dot{x}_4 = p_{11}(x_3 - x_4) - p_{12}p_{78}$$

$$\dot{x}_5 = p_{13}(p_{14}x_3 + p_{15}x_4 - p_{16}x_5 + (p_{59}x_{18}(p_{60}tanh(p_{61}x_{19}) - x_{17}) \times$$
$$(p_{62} - p_{63}tanh(p_{64}((x_5/p_{86}) - p_{65})))) -$$
$$(p_{69}x_{18}(p_{66} + p_{66}tanh(p_{67}((x_5/p_{86}) - p_{68})))))$$

$$\dot{x}_6 = p_{17}(x_3 - x_6) - p_{18}(p_{70} + p_{70}tanh(p_{71}(x_6 - p_{72})))$$

$$\dot{x}_7 = p_{19}(x_3 - x_7) - p_{20}(x_7 - x_8)$$

$$\dot{x}_8 = p_{21}(x_7 - x_8) - p_{22}(p_{73}(x_8/p_{87})(p_{73} + p_{74}tanh(p_{75}((x_{15}/p_{88}) - p_{68}))))$$

$$\dot{x}_9 = p_{23}(x_{10} - x_9)$$

$$\dot{x}_{10} = p_{24}(p_{25}x_9 + p_{26}x_{12} + p_{27}x_{13} + p_{28}x_{14} - p_{19}x_{10})$$

$$\dot{x}_{11} = p_{30}(x_{10} - x_{14}) - p_{31}(x_{14} - x_{15})$$

$$\dot{x}_{12} = p_{32}(x_{10} - x_{11})$$

$$\dot{x}_{13} = p_{33}(p_{34}x_{10} + p_{35}x_{11} - p_{36}x_{12} - (p_{91}(p_{34}x_{10} + p_{35}x_{11})))$$

$$\dot{x}_{14} = p_{37}(x_{10} - x_{13}) - p_{38}(p_{92}(p_{27}x_{13}))$$

$$\dot{x}_{15} = p_{39}(x_{14} - x_{15}) - p_{40}((x_{15})/((p_{90} - p_{93})/p_{93})(p_{57}) - (p_{58}))$$

$$\dot{x}_{16} = p_{41}(p_{42} - p_{43}tanh(p_{44}((x_{12}/p_{45}) - p_{46})) - x_{16})$$

$$\dot{x}_{17} = p_{47}(((p_{48}tanh(p_{49}x_{19}) - p_{50})/p_{51}) - x_{17})$$

$$\dot{x}_{18} = p_{52}(p_{53}tanh((p_{54}x_{12})/p_{55}) - x_{18})$$

$$\dot{x}_{19} = p_{56}(((p_{79} - p_{80}tanh(p_{81}((x_3/p_{88}) - p_{83}))) \times$$
$$(p_{82} - p_{83}tanh(p_{84}((x_{10}/p_{89}) - p_{85})))) - x_{19})$$

$$\text{(4.3)}$$

where $x_1 = G_{BV}$, $x_2 = G_{BI}$, $x_3 = G_H$, $x_4 = G_G$, $x_5 = G_L$, $x_6 = G_K$, $x_7 = G_{PV}$, $x_8 = G_{PI}$, $x_9 = I_B$, $x_{10} = I_H$, $x_{11} = I_G$, $x_{12} = I_L$, $x_{13} = I_K$, $x_{14} = I_{PV}$, $x_{15} = I_{PI}$, $x_{16} = M^I_{HGP}$, $x_{17} = M^I_{HGU}$, $x_{18} = f_2$, $x_{19} = G_C$.

4.2 Evolutionary optimization algorithms for parameter estimation

Evolutionary algorithms are effective in solving optimization problems. Particularly with the parameter estimation problem, biological models are difficult to solve with traditional estimation techniques [4]. Therefore, evolutionary algorithms have been frequently applied to this problem. This chapter provides the results of parametric estimation using the bioinspired strategies: evolutionary and particle swarm-based algorithms.

4.2.1 Methodology for parametric estimation

The set of parameters to be estimated are selected according to the sensitivity analysis of the Sorensen and Dalla Man models, as published in [5,6]. It is observed that the dynamics of blood glucose is complex, due to this, it is difficult to faithfully reproduce said dynamics with fixed parameters. Furthermore, the information that can be provided to the model is limited. This is why a time window is chosen, the Dalla Man and Sorensen model of 25 min. In this period, the models are capable of reproducing the dynamics with a good approximation to the real data of a patient with DM1. The parameter estimation for the three algorithms is summarized in the following steps:

Algorithm 5

1: Set interval $[t_i, t_f]$
2: Extract a time window from t_i to t_f from the collected data $X_D = [x_g, x_m, x_i]$ (glucose, carbohydrates and insulin).
3: Estimate the vector of unknown parameters p.
4: Solves the compartmental model in the initial time t_i until the final time t_f. The solution in the time window is considered in the output variable of the compartmental model that represents glucose levels via the interstitial pathway $\hat{G}_i$.
5: Calculate the error f_e using (Eq. (4.4)):

$$f_e = \frac{1}{n} \sum_{m=t_i}^{t_f} \sqrt{(log(\hat{G}_i(m)) - log(x_g(m)))^2} \tag{4.4}$$

where n represents the total data used for the evaluation in the time window. $\hat{G}_i(m)$ represents the glucose concentration at each time m from t_i to t_f. $x_g(m)$ stores the actual blood glucose concentration at time m.
6: If the error cannot be reduced, go to the next step, otherwise go back to step (3).
7: The time window $t_i \leftarrow t_i + 25 \, t_f \leftarrow t_f + 25$ is updated.
8: If there is more data in the next time window, go to step (2) otherwise finish.

4.3 Parametric estimation results in compartmental models

4.3.1 Sorensen parametric estimation results

To simulate Sorensen's physiological model, Γ_{meal} and $i(t)$ are considered as input to the system, where Γ_{meal} represents the total glucose absorbed by the patient's intestine due to meals and $i(t)$ is the insulin delivered to the patient subcutaneously via an insulin pump.

$y = G_{PI} = x_8$ is the output system representing glucose in the periphery interstitial fluid space.

Numerical experiments have been performed to estimate the glucose concentration in the interstitial space of peripheral tissue, G_{PI}, from parameter sets calculated by evolutionary and particle swarm-based algorithms. The experiment with each algorithm is represented in the following subsections.

4.3.1.1 Results obtained by Evonorm
The results of the experiments obtained by Evonorm are shown in Fig. 4.2. The corresponding real data (interstitial DM1 patient measurement value) is shown in red, while in blue the estimated data for the Evonorm algorithm are represented, through the parameters established in the corresponding window. The total mean error of the experiment is 10.9146 mg/dl with a standard deviation of 3.3052 mg/dl, as displayed in Table 4.2.

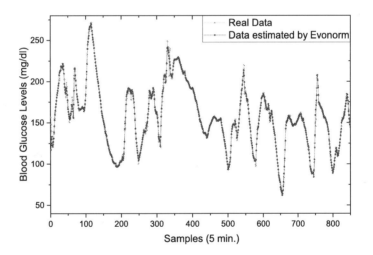

FIGURE 4.2 Blood glucose concentration recorded by the CGM. The glucose concentration estimated by the Evonorm algorithm.

Table 4.1 shows the mean and standard deviation of each parameter identified by the algorithm through the time windows. It is observed that the Evonorm algorithm achieves a small standard deviation in the estimated parameters, which means that it maintains a small range of variation in the parameters over time.

4.3.1.2 Results obtained by Differential Evolution
The results of the experiments obtained by Differential Evolution are shown in Fig. 4.3. The corresponding real data (interstitial DM1 patient measurement value) is shown in red, while in blue the estimated data for the DE algorithm are represented, through the parameters established in the corresponding window. The total mean error of the experiment is 11.9856 mg/dl with a standard deviation of 3.4640 mg/dl, as displayed in Table 4.4.

Table 4.1   Average and standard deviation of the parameters estimated by the Evonorm algorithm.

	Evonorm	
Parameter	Mean	SD
p_{61}	1.1519	0.5225
p_{62}	1.5365	0.5186
p_{63}	2.2312	0.7234
p_{64}	1.4508	0.4406
p_{65}	0.6346	0.5111
p_{73}	3.1953	1.4694
p_{74}	0.5279	0.3035
p_{75}	1.2042	0.7672
p_{80}	5.5485	1.3074
p_{82}	1.3985	0.2646

Table 4.2   The mean square error (MSE) and the standard deviation (SD) obtained by the Evonorm algorithm.

Algorithm	MSE (mg/dl)	SD (mg/dl)
Evonorm	10.9146	3.3052

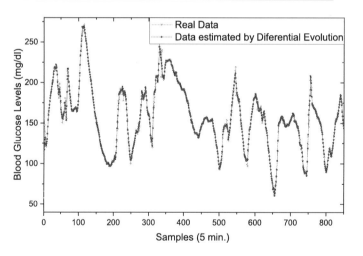

FIGURE 4.3 Blood glucose concentration recorded by the CGM. The glucose concentration estimated by the DE algorithm.

Table 4.3 shows the mean and standard deviation of each parameter identified by the DE algorithm through the time windows. It is observed that the algorithm achieves a slightly higher standard deviation than the Evonorm evolutionary algorithm. However, it shows similar results, so the values of the parameters coincide between the different evolutionary strategies.

Table 4.3 Average and standard deviation of the parameters estimated by the DE algorithm.

Parameter	DE	
	Mean	SD
p_{61}	1.0991	0.7494
p_{62}	1.5879	0.7563
p_{63}	2.2919	0.9731
p_{64}	1.3428	0.6264
p_{65}	0.7825	0.7520
p_{73}	3.7678	1.6868
p_{74}	0.5336	0.3891
p_{75}	1.3673	1.0263
p_{80}	5.2662	1.6304
p_{82}	1.4087	0.3512

Table 4.4 The mean square error (MSE) and the standard deviation (SD) obtained by the DE algorithm.

Algorithm	MSE (mg/dl)	SD (mg/dl)
Differential Evolution	11.9856	3.4640

4.3.1.3 Results obtained by PSO

The results of the experiments obtained by PSO are shown in Fig. 4.4. The corresponding real data (interstitial DM1 patient measurement value) is shown in red, while in blue the estimated data for the PSO algorithm are represented, through the parameters established in the corresponding window. The total mean error of the experiment is 10.9146 mg/dl with a standard deviation of 3.3052 mg/dl, as displayed in Table 4.6.

Table 4.5 shows the mean and standard deviation of each parameter identified by the PSO algorithm through the time windows. It is observed that the algorithm achieves a higher standard deviation than that shown by evolutionary algorithms. The results shown by the PSO algorithm show a mean square error similar to the evolutionary algorithms, so its performance does not differ greatly.

4.3.1.4 Results obtained by ACO

The results of the experiments obtained by ACO are shown in Fig. 4.5. The corresponding real data (interstitial DM1 patient measurement value) is shown in red, while in blue the estimated data for the ACO algorithm are represented, through the parameters established in the corresponding window. The total mean error of the experiment is 12.8714 mg/dl with a standard deviation of 3.5891 mg/dl, as displayed in Table 4.8.

Table 4.7 shows the mean and standard deviation of each parameter identified by the ACO algorithm through the time windows. In the case of the ant-based algorithm, it shows

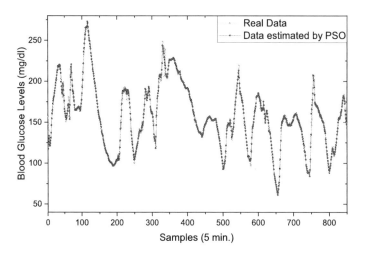

FIGURE 4.4 Blood glucose concentration recorded by the CGM. The glucose concentration estimated by the PSO algorithm.

Table 4.5   Average and standard deviation of the parameters estimated by the PSO algorithm.

Parameter	PSO	
	Mean	SD
P_{61}	0.9956	0.7499
P_{62}	1.6685	0.8258
P_{63}	2.2476	1.0406
P_{64}	1.2720	0.6122
P_{65}	0.7901	0.7832
P_{73}	3.8078	1.8079
P_{74}	0.5174	0.4014
P_{75}	1.5166	1.0306
P_{80}	5.1981	1.7151
P_{82}	1.3578	0.3598

Table 4.6   The mean square error (MSE) and the standard deviation (SD) obtained by the PSO algorithm.

Algorithm	MSE (mg/dl)	SD (mg/dl)
PSO	10.984	3.3159

a standard deviation in the parameters, which is why it almost always converges to them; however, it presents the highest mean square error of all the proposed algorithms.

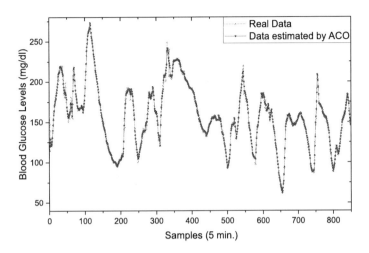

FIGURE 4.5 Blood glucose concentration recorded by the CGM. The glucose concentration estimated by the ACO algorithm.

Table 4.7 Average and standard deviation of the parameters estimated by the ACO algorithm.

	ACO	
Parameter	**Mean**	**SD**
P_{61}	1.0990	0.5171
P_{62}	1.5940	0.4664
P_{63}	2.2142	0.6635
P_{64}	1.3618	0.3908
P_{65}	0.7216	0.4347
P_{73}	3.7087	0.9885
P_{74}	0.4541	0.2851
P_{75}	1.3985	0.6959
P_{80}	5.4343	1.0556
P_{82}	1.3909	0.2321

Table 4.8 The mean square error (MSE) and the standard deviation (SD) obtained by the ACO algorithm.

Algorithm	MSE (mg/dl)	SD (mg/dl)
ACO	12.8714	3.5891

4.3.2 Dalla Man parametric estimation results

To simulate Dalla Man's physiological model, D and $u(t)$ are considered as the input of the system. D represents the total carbohydrate eaten at each meal, and $u(t)$ is the insulin administered subcutaneously. $y = SG(t)$ is the output system representing the glucose obtained by the sensor.

Numerical experiments were performed to estimate sensor glucose SG, from the parameter sets calculated by the evolutionary and particle swarm-based algorithms in the following subsections.

4.3.2.1 Results obtained by Evonorm

The results of the experiments obtained by Evonorm are shown in Fig. 4.6. The corresponding real data (interstitial DM1 patient measurement value) is shown in red, while in blue the estimated data for the Evonorm algorithm are represented, through the parameters established in the corresponding window. The total mean error of the experiment is 3.87351 mg/dl with a standard deviation of 2.10814 mg/dl, as displayed in Table 4.10.

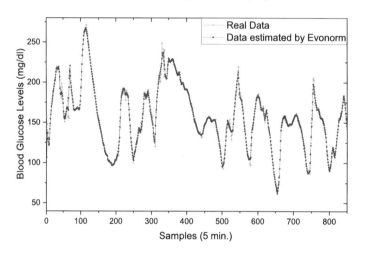

FIGURE 4.6 Blood glucose concentration recorded by the CGM. The glucose concentration estimated by the Evonorm algorithm.

Table 4.9 shows the mean and standard deviation of each parameter identified by the Evonorm algorithm through the time windows. It is observed that the parameters in the Dalla Man model vary slightly due to the complex dynamics of the real data, however, the identification results are better than those shown with the Sorensen model according to the identification error, so the Evonorm algorithm shows better performance in parametric estimation in the Dalla Man Model.

4.3.2.2 Results obtained by DE

The results of the experiments obtained by DE are shown in Fig. 4.7. The corresponding real data (interstitial DM1 patient measurement value) is shown in red, while in blue the estimated data for the DE algorithm are represented, through the parameters established in the corresponding window. The total mean error of the experiment is 4.05 mg/dl with a standard deviation of 2.3162 mg/dl, as displayed in Table 4.12.

Table 4.9   Average and standard deviation of the parameters estimated by the Evonorm algorithm.

Parameter	Evonorm	
	Mean	SD
k_p2	0.0460	0.0546
k_1	0.0203	0.0286
k_2	0.0057	0.0175
k_p1	0.9803	1.1033
k_i	0.0022	0.0027
k_abs	0.0011	0.0046
k_e1	2.87e-04	3.1e-04
k_{max}	0.0227	0.0274
k_{min}	0.0027	0.0029
k_p3	0.0026	0.0028
k_{gri}	0.0216	0.0269

Table 4.10   The mean square error (MSE) and the standard deviation (SD) obtained by the Evonorm algorithm.

Algorithm	MSE (mg/dl)	SD (mg/dl)
Evonorm	3.87351	2.10814

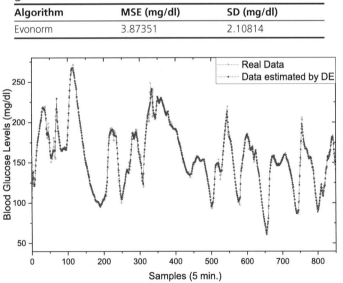

FIGURE 4.7 Blood glucose concentration recorded by the CGM. The glucose concentration estimated by the DE algorithm.

Table 4.11 shows the mean and standard deviation of each parameter identified by the DE algorithm through the time windows. Differential evolution shows a range of parameter values similar to those obtained by PSO, but the identification results are slightly inferior in performance. However, it does not differ too much from the results obtained with respect

Table 4.11 Average and standard deviation of the parameters estimated by the DE algorithm.

Parameter	DE	
	Mean	SD
k_p2	0.0633	0.0568
k_1	0.0278	0.0294
k_2	0.0073	0.0210
k_p1	1.3724	1.1964
k_i	0.0031	0.0029
k_abs	0.0014	0.0035
k_e1	3.25e-04	2.95e-04
k_{max}	0.0284	0.0303
k_{min}	0.0033	0.0029
k_p3	0.0031	0.0029
k_{gri}	0.0296	0.0296

to algorithms based on particle swarms. Furthermore, the results continue to be superior to those found by differential evolution in the Sorensen model.

Table 4.12 The mean square error (MSE) and the standard deviation (SD) obtained by the DE algorithm.

Algorithm	MSE (mg/dl)	SD (mg/dl)
DE	4.05	2.3162

4.3.2.3 Results obtained by PSO

The results of the experiments obtained by PSO are shown in Fig. 4.8. The corresponding real data (interstitial DM1 patient measurement value) is shown in red, while in blue the estimated data for the PSO algorithm are represented, through the parameters established in the corresponding window. The total mean error of the experiment is 3.76369 mg/dl with a standard deviation of 1.98821 mg/dl, as displayed in Table 4.14.

Table 4.13 shows the mean and standard deviation of each parameter identified by the PSO algorithm through the time windows. It is observed that the values of the parameters with respect to those found by the Evonorm algorithm are not similar, however, PSO shows better identification performance, achieving a lower error.

4.3.2.4 Results obtained by ACO

The results of the experiments obtained by ACO are shown in Fig. 4.9. The corresponding real data (interstitial DM1 patient measurement value) is shown in red, while in blue the estimated data for the ACO algorithm are represented, through the parameters established in the corresponding window. The total mean error of the experiment is 5.92666 mg/dl with a standard deviation of 3.92342 mg/dl, as displayed in Table 4.15.

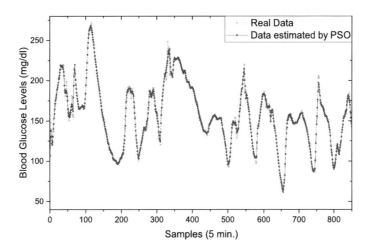

FIGURE 4.8 Blood glucose concentration recorded by the CGM. The glucose concentration estimated by the PSO algorithm.

Table 4.13 Average and standard deviation of the parameters estimated by the PSO algorithm.

	PSO	
Parameter	Mean	SD
k_p2	0.0689	0.0545
k_1	0.0371	0.0313
k_2	0.0082	0.0200
k_p1	1.5790	1.2023
k_i	0.0035	0.0026
k_abs	0.0131	0.0227
k_e1	3.28e-04	2.6e-04
k_{max}	0.0338	0.0284
k_{min}	0.0034	0.0028
k_p3	0.0034	0.0028
k_{gri}	0.0341	0.0289

Table 4.14 The mean square error (MSE) and the standard deviation (SD) obtained by the PSO algorithm.

Algorithm	MSE (mg/dl)	SD (mg/dl)
PSO	3.76369	1.98821

Table 4.16 shows the mean and standard deviation of each parameter identified by the ACO algorithm through the time windows. Of all the algorithms used, ACO is the one that presents the worst identification results, showing a large standard deviation in the parametric estimate, which implies that it does not have good precision and that the pa-

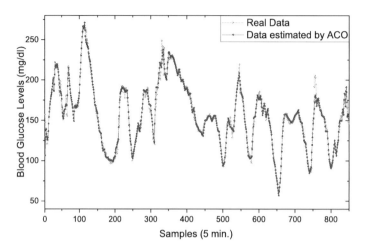

FIGURE 4.9 Blood glucose concentration recorded by the CGM. The glucose concentration estimated by the ACO algorithm.

Table 4.15   Average and standard deviation of the parameters estimated by the ACO algorithm.

Parameter	ACO	
	Mean	SD
k_p2	0.0671	0.0653
k_1	0.0223	0.0268
k_2	0.0048	0.0147
k_p1	1.3917	1.2538
k_i	0.0033	0.0031
k_abs	0.0021	0.0074
k_e1	3.29e-04	2.97e-04
k_{max}	0.0323	0.0304
k_{min}	0.0030	0.0030
k_p3	0.0031	0.0030
k_{gri}	0.0297	0.0296

Table 4.16   The mean square error (MSE) and the standard deviation (SD) obtained by the ACO algorithm.

Algorithm	MSE (mg/dl)	SD (mg/dl)
ACO	5.92666	3.92342

rameters vary greatly throughout the real data. ACO shows improvement with respect to the error in the results shown in the Sorensen model.

References

[1] Chis Oana-Teodora, Julio R. Banga, Eva Balsa-Canto, Structural identifiability of systems biology models: a critical comparison of methods, PLoS ONE 6 (11) (2011) e27755.

[2] Ahmad Haidar, Malgorzata E. Wilinska, James A. Graveston, Roman Hovorka, Stochastic virtual population of subjects with type 1 diabetes for the assessment of closed-loop glucose controllers, IEEE Transactions on Biomedical Engineering 60 (12) (2013) 3524–3533.

[3] Roberto Visentin, Chiara Dalla Man, Claudio Cobelli, One-day Bayesian cloning of type 1 diabetes subjects: toward a single-day UVA/Padova type 1 diabetes simulator, IEEE Transactions on Biomedical Engineering 63 (11) (2016) 2416–2424.

[4] Choujun Zhan, Wuchao Situ, Lam Fat Yeung, Peter Wai-Ming Tsang, Genke Yang, A parameter estimation method for biological systems modelled by ODE/DDE models using spline approximation and differential evolution algorithm, IEEE/ACM Transactions on Computational Biology and Bioinformatics 11 (6) (2014) 1066–1076.

[5] Oscar D. Sánchez, Eduardo Ruiz-Velázquez, Alma Y. Alanís, Griselda Quiroz, Luis Torres-Treviño, Parameter estimation of a meal glucose–insulin model for TIDM patients from therapy historical data, IET Systems Biology 13 (1) (2019) 8–15.

[6] G. Quiroz, R. Femat, On hyperglicemic glucose basal levels in type 1 diabetes mellitus from dynamic analysis, Mathematical Biosciences 210 (2) (2007) 554–575.

Neural model for glucose–insulin dynamics

5.1 Identification

The mathematical models obtained from systems play an essential role in modern science and technology because they are applied to obtain a better understanding of a process, verification of theoretical models, synthesis of control systems, signal prediction, optimization of the process behavior, and calculation of variables that are not directly measurable. In general, a mathematical model can be defined as a mathematical law that links system inputs (causes) with outputs (effects). In this way, the relationships of the system are modeled in mathematical structures such as differential equations or systems of differential equations.

Experimental analysis of a system that is unknown through input and output measurements is known as identification. The measurements are evaluated in the identification process that generates a mathematical model as a result. If the structure of the model is known a priori, parametric identification methods can be used. Otherwise, nonparametric identification procedures must be applied, the result of the identification being an experimental model.

There are different definitions for the identification of systems, on the one hand, [1] defines identification as the determination based on input and output data observed in a specific system in which it is equivalent to the identified system. However, it is almost impossible to find a model that completely matches the physical plant. in fact, the input–output of the system contains noise, so the identified model is only an approximation of the plant. According to [2] the identification of the system describes the essential characteristics of the objective system, also the model must be expressed in a useful form, suitable for applications. Also, [3] defines four entities in the identification process: experimentation, selection of the model structure, model estimation, and validation, see Fig. 5.1.

Experimentation: during the experimentation, descriptive data of the behavior of the system in the operating range are collected. The purpose is to vary the input and observe the impact it has on the output.

Model structure selection: In this stage, the model structure is chosen from a set of candidate models. Then, the model parameters are estimated for a specific use.

Model estimation: After the selection of the candidate model, a particular model is chosen.

Validation: Once the model has been trained, it must be evaluated to see if the model has the characteristics necessary for its acceptance or rejection.

https://doi.org/10.1016/B978-0-44-322341-9.00014-8

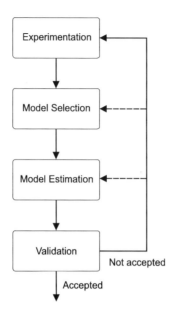

FIGURE 5.1 Basic system identification process.

The identification consists of selecting the model that best describes the data from the set of models built according to the design characteristics. In this way, the data, the model and specific criteria directly influence the identification performance, convergence precision, robustness, and computational complexity of the identification algorithm.

5.2 Identification with artificial neural networks

Obtaining a precise and faithful mathematical model is not an easy task due to its physically complex structures and hidden parameters [4]. Another way to derive a model is to obtain it from a data set collected through experimental data, which are often obtained through the stimulus–response or input–output strategy.

These identification methods can be a useful shortcut for the mathematical modeling of complex systems. However, system identification does not always result in an accurate model when describing the phenomenology, but it is possible to obtain a satisfactory model with reasonable effort, mainly for control or prediction tasks. The main problem is the requirement to carry out practical, realistic, and reproducible experiments to bring the system to its operational range [5].

In the literature, there are publications on the identification of glucose–insulin dynamics, most of them based on the linear approximation such as the ARX (AutoRegressive eXternal) [6], ARMA or ARMAX (AutoRegressive Moving Average with eXogenous inputs) [7] models. However, diabetes mellitus has a nonlinear behavior that is difficult to model by linear techniques, so nonlinear techniques such as neural networks are an alternative to determining mathematical models [8].

In addition, neural networks have become a useful methodology for solving difficult engineering problems, such as applications of modeling and control of complex nonlinear systems. Among the most common structures are: pre-fed and recurrent or feedback networks [9], [10], [11]. Many of the known results for system identification are applicable to identification using neural networks [4].

Neural networks to model glucose–insulin dynamics have been considered in various works [8], [12], [13], [14]. The types and architectures are varied, from the simple feedforward [8], [13], [14], to more sophisticated structures such as neurofuzzy systems [12]. Most of these works focus on developing models without the purpose of control, so they propose a non-affine model with a complex neural network structure, its application is restrictive due to the need for a large amount of external data that must be obtained from the patient such as dose and type of insulin, amount and type of exercise, carbohydrate content for each meal, among others. Hence, this information is difficult to provide to the patient.

This work also focuses on the identification of the unknown nonlinear dynamics of the blood glucose level in response to food and insulin infusion for patients with DM1. Different neural network models are proposed in order to capture the nonlinear behavior of blood glucose metabolism. The procedure performed is the one shown in Fig. 5.2.

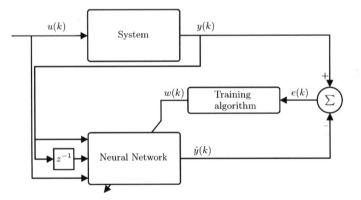

FIGURE 5.2 Architecture of glucose identification using neural networks.

Here, the unknown system is the DM1 patient, $u(k)$ is the input data, $y(k)$ is the output and is caused according to the unknown inputs and disturbances. $\hat{y}(k)$ is the output of the network model. $e(k)$ is the error produced by $y(k) - \hat{y}(k)$.

5.3 System description

The main reason for using NN is that it can be applied in a closed-loop system for online glucose concentration control. This requires collecting data from a glucose sensor in addition to the insulin dose administered by an insulin pump, as cited in [15]. The required data is collected using an insulin pump and a continuous glucose monitor.

Since the method mentioned above is an open-loop system, the amount of food is calculated by the patient under the guidance of the insulin pump manual or by a diabetes educator. Afterward, the insulin pump receives the carbohydrate intake. With this information, the device is programmed according to the patient's condition to determine the amount of insulin the pump should deliver. Additionally, continuous glucose monitoring receives a sample of interstitial blood glucose from the glucose sensor every five minutes. When the glucose sensor is linked to the patient, the user cannot access the data that the sensor provides for processing. In this case, the user is forced to wait until the sensor's service life ends.

5.3.1 Identification with artificial neural networks offline

The offline identification process with an artificial neural network involves training using historical or pre-existing data to capture patterns and relationships and then apply new data during the identification phase.

In the identification with deep neural networks, insulin samples $u_i(t)$, glucose $y(t)$, and carbohydrates $u_g(t)$ were used.

Below are the results obtained by the different neuronal models implemented for the identification, first the neuronal models were trained offline, these are the LSTM, BiLSTM, and MLP. For the neural network training, the Levenberg–Marquardt backpropagation algorithm and Adam's algorithm were used. Training data correspond to 80% of the data, while 20% was used for validation.

5.3.1.1 Neuronal network configuration using CGM

Neural networks have been trained to model glucose dynamics. That is, the neural networks are trained to generate a direct prediction of $y(k+1)$.

The information entering the neural networks is the current glucose measurement $y(k)$, and some past glucose samples measured by the CGM system. To produce the prediction, the three proposed NN models take into account past glucose information up to 40 min before the current time.

Since the sensor records glucose every five minutes, the neural network model needs eight sensor samples as inputs, which is equivalent to a 40-min memory. The neural networks were trained using sensor data as inputs in addition to carbohydrate and insulin intake, 80% of the data was used for training and 20% for validation.

Different configurations for neuronal networks were tested. The best structures for neural networks are described below.

For the LSTM neuronal network two hidden layers with 100 and 50 neurons were used, respectively. The same configuration was used for the BiLSTM neuronal network, with 100 neurons for the first layer and 50 for the second. The MLP neuronal network has a two-layer structure, 20 neurons for the first layer and 15 for the second layer. The weights and biases of the neuronal networks were randomly initialized and the training was carried out using the Levenberg–Marquardt retropropagation algorithm for MLP and the ADAM algorithm for LSTM and BiLSTM. Training is applied in batches.

5.3.1.2 LSTM identification results

The results of the identification are presented in Fig. 5.3. The LSTM is able to capture the dynamics of glucose, that is, the MSE obtained by the LSTM is 5.7816 (mg/dl) with a SD of 3.48329 (mg/dl), as shown in Table 5.1.

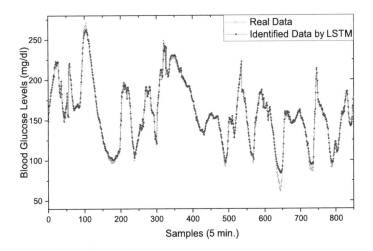

FIGURE 5.3 The patient's blood glucose concentration estimated using the LSTM network.

Table 5.1 The mean square error (MSE) and the standard deviation (SD) obtained by LSTM neural network.

Algorithm	MSE (mg/dl)	SD (mg/dl)
LSTM	5.7816	3.48329

It can be seen that although the glucose level varies from one moment to the next due to its nonlinear behavior, the LSTM neural network shows acceptable performance, showing a low identification error and a low standard deviation. It should be noted that the dynamics shown in the real data contain disturbances such as stress, exercise, and measurement errors by the glucose sensor and despite this the neural network captures the behaviors in the identification.

5.3.1.3 BiLSTM results

The results of the identification are presented in Fig. 5.4. The BiLSTM is able to capture the dynamics of glucose, that is, the MSE obtained by the BiLSTM is 5.7635 (mg/dl) with a SD of 3.73804 (mg/dl), as shown in Table 5.2.

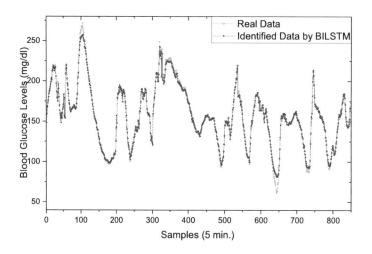

FIGURE 5.4 The patient's blood glucose concentration estimated using the BiLSTM network.

Table 5.2 The mean square error (MSE) and the standard deviation (SD) obtained by BiLSTM neural network.

Algorithm	MSE (mg/dl)	SD (mg/dl)
BiLSTM	5.7635	3.73804

According to the results obtained by the BiLSTM neural network, good identification results are obtained, quite similar to those obtained by the unidirectional LSTM neural network. Despite having a slightly better MSE, the standard deviation is greater than that of the LSTM. In general, it can be said that BiLSTM does not substantially improve identification results, but it is still an interesting tool for modeling tasks.

5.3.1.4 MLP results

The identification results are presented in Fig. 5.5. The MLP is able to capture the dynamics of glucose, that is, the MSE obtained by the MLP is 4.52478 (mg/dl) with an SD of 2.79436 (mg/dl), as shown in Table 5.3.

Table 5.3 The mean square error (MSE) and the standard deviation (SD) obtained by MLP neural network.

Algorithm	MSE (mg/dl)	SD (mg/dl)
MLP	4.52478	2.79436

For the results presented by the neural network by the MLP, it is observed that it has better identification performance than those obtained by the other two deep neural networks trained offline. This shows that a neural network with not-so-complex operations

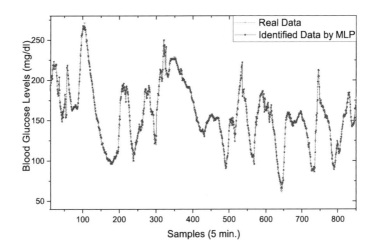

FIGURE 5.5 The patient's blood glucose concentration estimated using the MLP network.

is capable of modeling nonlinear dynamics such as that of glucose insulin in human patients with type 1 diabetes mellitus. Although the results are good, it should be noted that the behavior presented by the neural networks trained offline must be retrained to maintain performance, since the dynamics of glucose insulin varies with respect to time due to different situations that patients experience.

5.3.2 Identification with an artificial neural network online

Online training neural networks for identification refers to the process of continuously updating the neural network model as data is collected or as new data becomes available in real time. This process allows the model to adjust and improve in performance over time. It is particularly useful in dynamic environments where the distribution of data changes or varies over time.

5.3.2.1 RHONN identification structure

The following structure is proposed for the high-order recurrent neural network for the identification of glucose dynamics:

$$
\begin{aligned}
\hat{x}_1(k+1) &= w_{11}z_1^2 \\
\hat{x}_2(k+1) &= w_{21}(z_1^5z_2^4) + w_{22}z_1^5 + w_{23}z_2^3 + w_{12}u_c(k) + w_{13}u_i(k)
\end{aligned}
\tag{5.1}
$$

where $x_1 = y(k-1)$, $x_2 = y(k)$, $u_c(k) = carbohydrates$, $u_i(k) = insulin$, $z_1 = tanh(x_1)$, $z_2 = tanh(x_2)$. The parameters w_{11}, w_{12}, and w_{13} are fixed and are set to the values 2, 0.001, and 0.001, respectively. Then w_{21}, w_{22}, and w_{22} are adjustable weights, and k represents the sampling period of 5 min.

The scheme is illustrated in Fig. 5.6, the input signals to the RHONN set equations (5.1) are the glucose absorption and the insulin infusion ($u(k) = [u_c(k), u_i(k)]$), the glucose level measured by the sensor are state variables $y(k-1) = x_1$, $y(k) = x_2$.

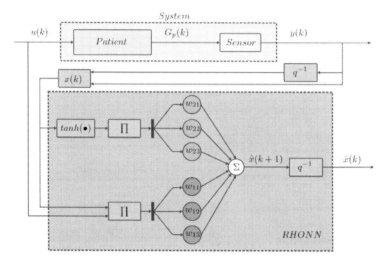

FIGURE 5.6 RHONN scheme for glucose identification. $G_p(k)$ represents the patient's blood glucose level in the sample k that does not contain noise or delays added by the sensor (not accessible).

The training is carried out by means of the EKF training algorithm with the following values: $R_i(0) = Q_i(0) = 1 \times 10^{-10}$ and $P_i(0) = 1$ on the main diagonal.

5.3.3 RHONN results

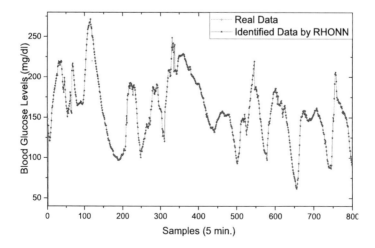

FIGURE 5.7 The patient's blood glucose concentration estimated using the RHONN neural network.

The identification results are shown in Fig. 5.7. As can be seen, the neural network captures the dynamics of glucose since it achieves an MSE of 8.10257 (mg/dl) with a standard deviation of 1.67833 (mg/dl).

The glucose concentration of a patient with type 1 diabetes mellitus varies throughout the day and from one day to the next caused by routine activities. Unlike neural networks trained offline, the RHONN neural network presents superior identification results, in addition to having the advantage that the neural network is updated for each sample.

References

[1] Lotfi A. Zadeh, From circuit theory to system theory, Proceedings of the IRE 50 (5) (1962) 856–865.

[2] P. Eykhoff, System Identification-Parameter and State Estimation, John Wiley & Sons, London, New York, Sydney, Toronto, 1974.

[3] Ljung Lennart, Convergence analysis of parametric identification methods, IEEE Transactions on Automatic Control 23 (5) (1978) 770–783.

[4] L. Ljung, System Identification – Theory for the User, 2nd edn., Prentice Hall, Upper Saddle River, NJ, 1999, p. 101, Feasible mind uploading.

[5] Jay A. Farrell, Marios M. Polycarpou, Adaptive Approximation Based Control: Unifying Neural, Fuzzy and Traditional Adaptive Approximation Approaches, John Wiley & Sons, 2006.

[6] Tom Van Herpe, Marcelo Espinoza, Bert Pluymers, Ivan Goethals, Pieter Wouters, Greet Van den Berghe, Bart De Moor, An adaptive input–output modeling approach for predicting the glycemia of critically ill patients, Physiological Measurement 27 (11) (2006) 1057.

[7] Meriyan Eren-Oruklu, Ali Cinar, Lauretta Quinn, Donald Smith, Adaptive control strategy for regulation of blood glucose levels in patients with type 1 diabetes, Journal of Process Control 19 (8) (2009) 1333–1346.

[8] A. Karim El-Jabali, Neural network modeling and control of type 1 diabetes mellitus, Bioprocess and Biosystems Engineering 27 (2005) 75–79.

[9] Alma Y. Alanis, Edgar N. Sanchez, Alexander G. Loukianov, Discrete-time adaptive backstepping nonlinear control via high-order neural networks, IEEE Transactions on Neural Networks 18 (4) (2007) 1185–1195.

[10] David C. Klonoff, The artificial pancreas: how sweet engineering will solve bitter problems, Journal of Diabetes Science and Technology 1 (1) (2007) 72–81.

[11] Sarah Wild, Gojka Roglic, Anders Green, Richard Sicree, Hilary King, Global prevalence of diabetes: estimates for the year 2000 and projections for 2030, Diabetes Care 27 (5) (2004) 1047–1053.

[12] Marcos A. González-Olvera, Ana G. Gallardo-Hernández, Yu Tang, Maria Cristina Revilla-Monsalve, Sergio Islas-Andrade, A discrete-time recurrent neurofuzzy network for black-box modeling of insulin dynamics in diabetic type-1 patients, International Journal of Neural Systems 20 (02) (2010) 149–158.

[13] Stavroula G. Mougiakakou, Aikaterini Prountzou, Dimitra Iliopoulou, Konstantina S. Nikita, Andriani Vazeou, Christos S. Bartsocas, Neural network based glucose-insulin metabolism models for children with type 1 diabetes, in: 2006 International Conference of the IEEE Engineering in Medicine and Biology Society, IEEE, 2006, pp. 3545–3548.

[14] W.A. Sandham, D.J. Hamilton, A. Japp, K. Patterson, Neural network and neuro-fuzzy systems for improving diabetes therapy 3 (1998) 1438–1441.

[15] Alma Y. Alanis, Blanca S. Leon, Edgar N. Sanchez, Eduardo Ruiz-Velazquez, Blood glucose level neural model for type 1 diabetes mellitus patients, International Journal of Neural Systems 21 (06) (2011) 491–504.

6

Multistep predictor applied to T1DM patients

6.1 Prediction

Multistep forward forecasting, also known as multistep forecasting or multistep time series forecasting, is a methodology used in statistics and machine learning to predict the values in the future of one or more variables in a series of time several instants of time forward.

The goal of traditional time series forecasting is to predict a future value of a variable. However, in the prediction of several steps forward it expands the concept to multiple values in the future time. Hence, instead of predicting a single value $k + 1$, it is possible to predict $k + n$, where n is the prediction horizon.

Knowledge of various values in the future of a time series can be very useful in various applications such as predicting stock prices, weather forecasting, demand forecasting in supply chain management, as well as securities futures, blood glucose levels, etc.

Making a several-step-ahead prediction can be more challenging than a one-step-ahead prediction because errors can accumulate over time, making it difficult to maintain accuracy over longer time horizons. Therefore, sophisticated models and techniques are often employed, such as recurrent neural networks (RNN), long short-term memory networks (LSTM), and other advanced time series forecasting methods.

6.1.1 Multistep ahead prediction strategies

The multistep ahead prediction consists of predicting the value $\hat{x}_{k+n}$, according to the time series $[x_1, x_2...x_k]$ with k observed data.

There are different prediction strategies, but the most used are recursive prediction, also called iterative or multi-stage, and independent or direct prediction.

- Recursive strategy: in this strategy a model is trained to predict one step ahead $(k + 1)$, then the predicted value plus the previous values are propagated to the next model that predicts one step forward, this strategy uses n models to forecast n steps ahead.
- Direct strategy: a single model is trained to generate a forward forecast for time $k + n$ [1], [2], [3].

Different works propose the use of the direct or recursive strategy; however, it is not clear which of them is the best. Both strategies propagate forecast error rapidly in long-term forecasts, causing forecast degradation. Both strategies are used in this book to show that neural models can be especially useful in long-term forecasting.

Bio-Inspired Strategies for Modeling and Detection in Diabetes Mellitus Treatment
https://doi.org/10.1016/B978-0-44-322341-9.00015-X

6.1.2 Recursive strategy

In the iterative or recursive strategy a single model is trained to map a data series $[x_1, x_2...x_k]$ through $\hat{f}$ to a one-step-ahead prediction [4], this is:

$$\hat{x}_k = \hat{f}(x_{k-1}, x_{k-2}, ..., x_{k-d}, u_{k-1}, ..., u_{k-d}) \tag{6.1}$$

where $1, 2, ...d$ are delays in the output and input. Then, the model $\hat{f}$ is used to predict the next value $\hat{x}_{k+1}$:

$$\hat{x}_{k+1} = \hat{f}(\hat{x}_k, x_{k-1}, ..., x_{k-d+1}, ..., \hat{u}_k, ..., u_{k-d+1}) \tag{6.2}$$

The result of forecasting one step ahead is used to predict the next steps by relating: $\hat{x}_{k+n}$ with $\hat{x}_{k+n-1}, \hat{x}_{k+n-2}, ..., x_k, x_{k-1}, ..., x, \hat{u}_{k+n-1}, \hat{u}_{k+n-2}, ..., u_k, u_{k-1}$. The $\hat{x}_{k+m}$ values are predicted iteratively from $m = 1$ to $m = n$.

$$\begin{aligned}\hat{x}_{k+n} = \hat{f}(\hat{x}_{k+n-1}, \\ \hat{x}_{k+n-2}, ..., x_k, ..., x_{k-d+1}, \\ \hat{u}_{k+n-1}, \hat{u}_{k+n-2}, ..., u_k, ..., u_{k-d+1})\end{aligned} \tag{6.3}$$

It should be noted that the recursive strategy is susceptible to error accumulation when the prediction horizon is too long or the serial data contains noise, because the error produced in the predictions is propagated to the following [4]. This can be further affected if the forecast horizon n is greater than the number of lags d, which would cause, at some point, the input values to the model becoming all forecasts and not actual observed data. Fig. 6.1 shows the graphical representation of the recursive prediction.

6.1.3 Direct strategy

The direct strategy aims to predict the value y_{k+n} according to the observed values $[x_1, x_2, ..., x_k]$ independently from the other forecasts [4]. in this way, n different trained $\hat{f}_i$ models are used to predict different forecast horizons, where:

$$\hat{x}_{k+i} = \hat{f}_i(x_k, ..., x_{k-d+1}, u_k, ..., u_{k-d+1}) + e \tag{6.4}$$

with $i \in \{1, ..., n\}$, d is the number of delays used to predict, and e is the error, disturbances and/or noise.

As this strategy uses a single model to predict values of n independently, it does not accumulate much error [4]. On the other hand, this strategy is computationally expensive because as many trained models are needed as there is a size in the prediction horizon [4]. Fig. 6.2 shows the graphical representation of the direct prediction.

6.1.4 Neuronal network configuration using CGM data

The aforementioned neural networks were implemented for the MSA prediction of glucose by direct and recursive strategies. Glucose is forecasted for the prediction horizons of 15,

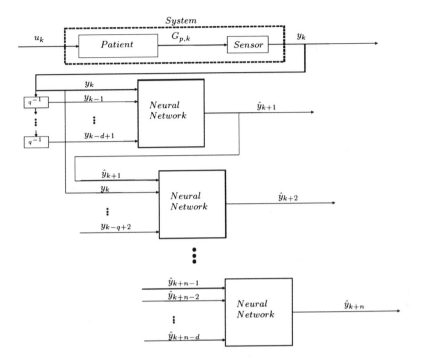

FIGURE 6.1 Recursive MSA prediction architecture using CGM data. $G_{p,k}$ is the patient's blood glucose level.

30, 45, and 60 min ahead, to compare the performance of the three proposed networks with the two different strategies.

For the case of the recursive strategy the MLP, LSTM, and BiLSTM are trained one step ahead, then it iteratively predicts $(\hat{y}_{k+1}, ..., \hat{y}_{k+n})$. On the other hand, in the direct strategy, the NN are trained to predict glucose at $n = 15, 30, 45$, and 60 min ahead. That is, a direct prediction y_{k+n} is generated for each prediction horizon.

Neural networks have as input the current glucose measurement y_k and some past glucose observations of up to 40 min to predict $\hat{y}_{k+n}$, with the three neural models and both strategies. Fig. 6.2 and Fig. 6.1 show the configuration of the NNs using online CGM data for the direct and recursive strategy, respectively. All models were trained using only sensor data as inputs. The CGM records glucose every 5 min, so that the input to the neural networks is equivalent to information from up to eight memory samples (40 min). Only 60% of the data was used for training and 40% for validation.

6.2 Prediction results evaluation criteria

To compare the performance of the neural networks and the methods, the following evaluation criteria (EC) were used: (1) mean square error (mg/dl), (2) correlation coefficient

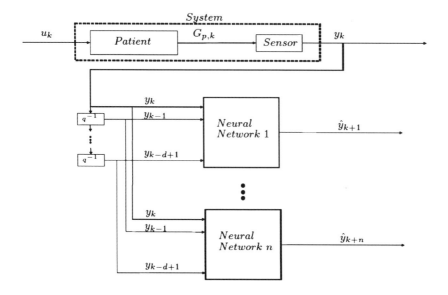

FIGURE 6.2 Direct MSA prediction architecture using CGM data. $G_{p,k}$ is the patient's blood glucose level.

(CC) and (3) time lag (TL, min). ECs indicate the ability of the models to predict, by comparing the observed data (y) and the predicted values ($\hat{y}$). RMSE indicates the prediction performance through the difference between y and $\hat{y}$. Then, the prediction performance is better the smaller the error. The correlation coefficient is used to determine the linear dependence between the observed and predicted data, the closer to 1, the greater the similarity between both data sets, the correlation coefficients used in this work are presented below [5]:

CC is the linear dependence between real data and predicted data, it is calculated by:

$$CC = \frac{\sum (y - y_{mean})(\hat{y} - \hat{y}_{mean})}{\sqrt{\sum (y - y_{mean})^2}\sqrt{\sum (\hat{y} - \hat{y}_{mean})^2}} \tag{6.5}$$

Time lag (TL) is the time interval in which the RMSE is smallest between the observed and predicted values. It is calculated by [6]:

$$T_\tau = \sqrt{\frac{1}{N - \tau} \sum_{i=1}^{N-\tau} (\hat{y}_i - y_{i+\tau})^2} \tag{6.6}$$

where τ is the time shift.

6.3 Results

6.3.1 Neural models with offline training

6.3.1.1 MLP results

Table 6.1 shows a summary of the results obtained from the MLP neural networks at different prediction horizons with the two strategies proposed in this chapter. Fig. 6.3 represents the prediction results with the different strategies for a prediction horizon of 30 min.

Table 6.1 MLP neural network results by the different strategies.

	Strategy	Direct	Recursive
	Prediction	**MLP**	**MLP**
15 min ahead	RMSE	13.77819	26.75611
	SD	7.50299	17.14318
	PCC	0.9630	0.8862599
	SCC	0.97562	0.92425
	Time Lag	2.5	2
30 min ahead	RMSE	19.85867	34.00122
	SD	12.83847	22.91996
	PCC	0.9141658	0.86977
	SCC	0.91482	0.9139132
	Time Lag	8	7.5
45 min ahead	RMSE	25.90918	36.51888
	SD	17.40887	19.76219
	PCC	0.8542224	0.8021991
	SCC	0.84553	0.808032
	Time Lag	17.5	17.5
60 min ahead	RMSE	32.54538	41.00433
	SD	21.13459	23.4473
	PCC	0.7570724	0.72951
	SCC	0.75221	0.7239
	Time Lag	22.5	27.5

According to the prediction results for a prediction horizon of 15 min, the RMSE of the direct and recursive strategy in the MLP are 13.77819 mg/dl and 26.75611 mg/dl, respectively. The lowest standard deviation is presented by the direct strategy of 7.50299 mg/dl. In addition, the correlation is better in the direct strategy for the MLP neural network. Finally, Time Lag is better in the recursive strategy.

For the prediction horizon of 30 min, the best RMSE is from the direct strategy of 19.85867 mg/dl with a standard deviation of 12.83847 mg/dl. The correlation presented is better in the direct strategy in addition to Time Lag.

For the prediction horizons of 45 and 60 min, it is shown that the error for the recursive method has been propagated, so the RMSE is considerably higher than that obtained by

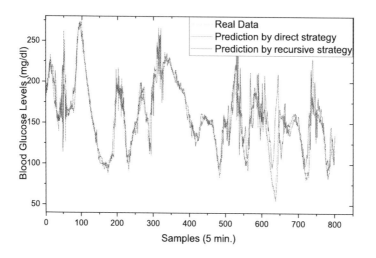

FIGURE 6.3 Results of the MLP NN for prediction 30 min ahead.

the direct method. In general, the MLP neural network shows a high RMSE at high prediction horizons. The best results were mostly presented by the direct strategy for the MLP neural network.

6.3.1.2 LSTM results

Table 6.2 shows a summary of the results obtained from the MLP neural networks at different prediction horizons with the two strategies proposed in this chapter. Fig. 6.4 represents the prediction results with the different strategies.

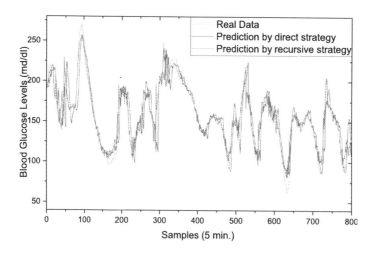

FIGURE 6.4 Results of the LSTM NN for prediction 30 min ahead.

Table 6.2 LSTM neural network results by the different strategies.

	Strategy	Direct	Recursive
	Prediction	LSTM	LSTM
15 min ahead	RMSE	12.09038	11.66631
	SD	8.28194	6.65180
	PCC	0.9717	0.9781522
	SCC	0.97685	0.989896
	Time Lag	7.5	2
30 min ahead	RMSE	17.22558	18.48139
	SD	12.37488	12.10129
	PCC	0.934922	0.9123528
	SCC	0.92258	0.946548
	Time Lag	12.5	12.5
45 min ahead	RMSE	24.55082	25.10456
	SD	17.95506	18.09081
	PCC	0.8627064	0.8455
	SCC	0.84089	0.8899326
	Time Lag	27.5	22.5
60 min ahead	RMSE	30.79154	31.44521
	SD	23.04362	23.45088
	PCC	0.7740924	0.79439
	SCC	0.75333	0.844297
	Time Lag	37.5	27.5

According to the prediction results for a prediction horizon of 15 min, the RMSE of the direct and recursive strategy in the LSTM are 12.09038 mg/dl and 11.66631 mg/dl, respectively. It can be seen that the recursive strategy has better prediction performance at low prediction horizons.

The lowest standard deviation is presented by the direct strategy of 6.65180 mg/dl. In addition, the correlation and Time Lag are better in the direct strategy for the LSTM neural network.

For the prediction horizon of 30 min, the best RMSE is from the direct strategy of 17.22558 mg/dl with a standard deviation of 12.37488 mg/dl. The correlation presented is better in the direct strategy, in addition, Time Lag is similar in both strategies.

For the prediction horizons of 45 and 60 min, it is shown that the error and standard deviation for the recursive method and the direct method are quite similar. In general, the results presented by the LSTM are good for both methods. It is worth mentioning that the lowest Time Lag is obtained by the recursive method of the LSTM.

6.3.1.3 BiLSTM results

Table 6.3 shows a summary of the results obtained from the BiLSTM neural networks at different prediction horizons with the two strategies proposed in this Chapter. Fig. 6.5

represents the prediction results with the different strategies at a prediction horizon of 30 min.

Table 6.3 BiLSTM neural network results by the different strategies.

	Strategy	Direct	Recursive
	Prediction	BiLSTM	BiLSTM
15 min ahead	RMSE	11.27294	12.97637
	SD	7.76026	8.21314
	PCC	0.9730	0.9698
	SCC	0.97759	0.9755
	Time Lag	7.5	2
30 min ahead	RMSE	17.3545	24.33325
	SD	12.39668	15.41082
	PCC	0.9355105	0.9059
	SCC	0.92289	0.9121
	Time Lag	12.5	12.5
45 min ahead	RMSE	24.65796	35.0736
	SD	17.95185	21.1327
	PCC	0.8626238	0.8220
	SCC	0.84377	0.84553
	Time Lag	27.5	22.5
60 min ahead	RMSE	30.61593	45.19504
	SD	22.76999	26.03341
	PCC	0.7769292	0.7220
	SCC	0.7546	0.7361
	Time Lag	37.5	27.5

According to the prediction results for a prediction horizon of 15 min, the RMSE of the direct and recursive strategy in the BiLSTM are 11.27294 mg/dl and 12.97637 mg/dl, respectively. The RMSE of the direct strategy is the lowest of all neural networks

The lowest standard deviation is presented by the direct strategy of 7.76026 mg/dl. In addition, the correlation is better in the direct strategy but the Time Lag presents better results in the recursive strategy for the BiLSTM neural network.

For the prediction horizon of 30 min, the best RMSE is from the direct strategy of 17.35453 mg/dl with a standard deviation of 12.39668 mg/dl. The correlation presented is better in the direct strategy, in addition, Time Lag is similar in both strategies.

For the prediction horizons of 45 and 60 min, it is shown that the error and standard deviation for the recursive method have increased significantly compared to the direct method. In general, the results presented by the BiLSTM are better for the direct method.

6.3.1.4 Results discussion

The results show that prediction 15 min ahead for the three proposed neural networks show similar results in both strategies, that is, approximately RMSE = 12 mg/dl and SD =

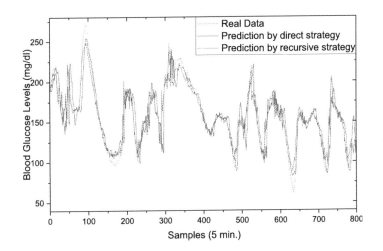

FIGURE 6.5 Results of the BiLSTM NN for prediction 30 min ahead.

8 mg/dl, with the exception of the MLP with recursive strategy that did not obtain a good RMSE. As mentioned above, this time is approximately the delay that exists between blood glucose and interstitial glucose, so that in either of the two methods, blood glucose can be estimated without much error. These results would help improve glucose control, however, it is even more valuable to forecast at a longer prediction horizon. The prediction 30 to 45 min ahead is an adequate prediction horizon to prevent risk scenarios (hypoglycemia and hyperglycemia), at this time the error is already accumulated for both strategies, being slightly higher for the recursive strategy, being in most networks of approximately RMSE = 17.09 mg/dl and SD = 8.28 mg/dl to RMSE = 34 mg/dl and SD = 22.91 mg/dl. Finally, the prediction 60 min ahead shows an RMSE for the direct strategy from RMSE = 30.79 mg/dl and SD = 23.04 mg/dl to RMSE = 32.54 mg/dl and SD = 23.45 mg/dl with a time lag of 22.5–37.5 min, for the recursive strategy; RMSE = 31.44 mg/dl and SD = 23.45 mg/dl to RMSE = 41 mg/dl and SD = 23.44 mg/dl, with a time lag of 27.5 min. The lowest TL is presented by the MLP by direct strategy (22.5 min). However, the best CC is from the LSTM by recursive strategy.

6.3.2 Neural models with online training

6.3.2.1 RHONN configuration for MSA prediction

As observed in the results from previous neural networks, recursive methods present lower prediction performance compared to direct methods. However, online training could improve prediction results over long horizons. To implement the prediction with the RHONN we start from the identified model and then consider n neural networks identical to the one used for identification, each of them is trained individually. Then, to prevent the predictor from guessing close to the mean distribution, RHONN is trained with the error $(y(k) - \hat{y}_n(k))$ to produce a different output [7]. To do this, we consider the current output of the system $y(k)$ and the expected output for that instant in each of the RHONN models $(\hat{y}_1(k), \hat{y}_2(k), ..., \hat{y}_n(k))$. The output of each network generates the prediction one step ahead. In this way, to determine the value in $k + n$, n neural networks are necessary to generate it. The representative diagram of the implemented structure is shown in Fig. 6.6.

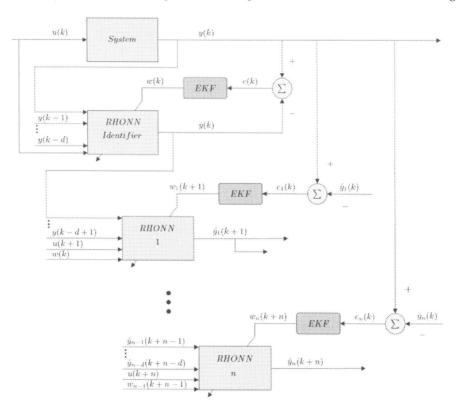

FIGURE 6.6 Multistep ahead prediction structure for RHONN.

6.3.3 RHONN results

To understand the performance of the models, evaluation criteria (EC) are used, which indicate the overall prediction ability by comparing the test data sets and the predicted value. The EC used were: root mean square error (mg/dl), correlation coefficient (CC), and time lag (TL, (min)). The results of the experiments are presented in Table 6.4. The prediction results of the 30-min prediction horizon are shown in Fig. 6.7.

For the 15-min advance prediction, the error and standard deviation are 16.7862 and 7.3944, respectively. The prediction 15 min ahead would roughly mean the delay that exists between blood glucose and interstitial glucose. This allows determination of the blood glucose, which helps the control algorithms calculate the insulin dose with higher pressure.

Table 6.4   RHONN neural network results by the different strategies.

Prediction	EC	15 min ahead	30 min ahead	45 min ahead	60 min ahead
	RMSE	16.7862	24.33317	30.60121	37.25601
	SD	7.3944	15.53141	21.49174	27.30774
RHONN	CC	0.9718	0.9538	0.9530	0.9523
	Time Lag	9.6542	12.6760	12.7989	12.9494

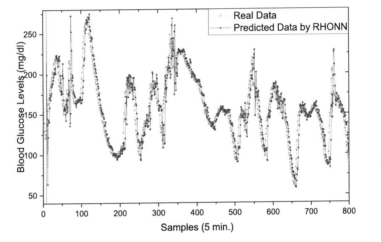

FIGURE 6.7 Glucose concentration recorded by the CGMS system and the corresponding glucose concentration predicted to 30 min ahead.

Diabetes has a nonlinear behavior, which causes the error at large prediction horizons to accumulate quickly. Online training is considered, so the network is updated as the data is obtained without prior knowledge. At a prediction horizon of 30 min the error is 24.33317 with a standard deviation of 15.5314.

For 45 and 60 min ahead the error has accumulated to 30.60121 and 37.25601, respectively. However, the time lag results are quite good, showing a value of approximately 13 min, in addition to showing a correlation of 0.95. The recursive strategy for the high-order recurrent neural network trained online shows satisfactory results at low computational cost.

The relevance of predicting events 30 min or more in advance lies in the ability to anticipate when glucose levels will exceed the risk limits, established at 300 mg/dl and 70 mg/dl. This makes it possible to prevent or at least reduce the time in which these events take place. Although most models seek to generally represent the behavior of glucose dynamics in a population, it is more effective to customize them for each individual. In this context, RHONN stands out by not requiring prior training and being implementable in real time in glucose sensors and mobile applications. This allows patients to know their current and future glucose levels up to 60 min later, facilitating preventive decision making. Additionally, RHONN provides a glucose model that can be used to develop a glucose control algorithm.

The RHONN results show that despite being trained online without prior knowledge of the data, it achieves a good performance, which in comparison to the other methods need to be pre-trained. The LSTM presented in this work agrees with the results obtained in [8]. This encourages us to use RHONN as an online predictor.

References

[1] Pin-An Chen, Li-Chiu Chang, Fi-John Chang, Reinforced recurrent neural networks for multi-step-ahead flood forecasts, Journal of Hydrology 497 (2013) 71–79.

[2] Junhwa Chi, Hyun-choel Kim, Prediction of arctic sea ice concentration using a fully data driven deep neural network, Remote Sensing 9 (12) (2017) 1305.

[3] Alma Y. Alanis, Oscar D. Sanchez, Jesus G. Alvarez, Time series forecasting for wind energy systems based on high order neural networks, Mathematics 9 (10) (2021) 1075.

[4] Souhaib Ben Taieb, Gianluca Bontempi, Amir F. Atiya, Antti Sorjamaa, A review and comparison of strategies for multi-step ahead time series forecasting based on the NN5 forecasting competition, Expert Systems with Applications 39 (8) (2012) 7067–7083.

[5] Mavuto M. Mukaka, A guide to appropriate use of correlation coefficient in medical research, Malawi Medical Journal 24 (3) (2012) 69–71.

[6] Andrew J. Conway, Keith P. Macpherson, John C. Brown, Delayed time series predictions with neural networks, Neurocomputing 18 (1–3) (1998) 81–89.

[7] Paulo Cortez, Pedro J. Pereira, Rui Mendes, Multi-step time series prediction intervals using neuroevolution, Neural Computing and Applications 32 (13) (2020) 8939–8953.

[8] Qingnan Sun, Marko V. Jankovic, Lia Bally, Stavroula G. Mougiakakou, Predicting blood glucose with an LSTM and Bi-LSTM based deep neural network, in: 2018 14th Symposium on Neural Networks and Applications (NEUREL), IEEE, 2018, pp. 1–5.

Classification and detection of diabetes mellitus and impaired glucose tolerance

7.1 Classification

Classification is a machine learning task whose objective is to predict the class or category through data processing. The classification can be binary, where the class only has two possibilities, or multiple classification, where the target variable can have more than two classes.

The most common algorithms in classification are Logistic Regression, Decision Trees, Random Forests, Support Vector Machines (SVM), K-Nearest Neighbors (KNN), and Neural Networks. The performance of the models is evaluated on a test data set using evaluation criteria such as accuracy, precision, recall, F1 score, and area under the ROC curve (AUC-ROC).

In general, classification is a fundamental task in machine learning. Among the main applications of classification are spam detection, natural language processing, medical diagnosis, image classification, object recognition, and credit risk assessment.

7.1.1 Time series classification

For its part, time series classification is a challenging task where the objective is to predict the class or category for a sequence of data. This serial data is acquired from real applications by means of sensors in real-time. This is why time series data confirms a trend since there is a relationship between past and present data. The difference with traditional classification is that the time dependence of the data is taken into account.

A time series can be defined as a vector $X = [x_1, x_2, ..., x_n]$ composed of real values, where n represents the size of the vector. In time series classification, a data set is defined as $D = \{(X_1, Y_1), (X_2, Y_2), ..., (X_N, Y_N)\}$, X_i represents the time series i, while the class label is represented by Y_i, where Y_i is a vector of length that depends on the number of classes k, the elements $j \in [1, q]$ are equal to 1 if X_i is j and 0 otherwise.

The data set D can then be used to train models with the ability to create a class probability distribution based on the space of possible inputs.

Bio-Inspired Strategies for Modeling and Detection in Diabetes Mellitus Treatment
https://doi.org/10.1016/B978-0-44-322341-9.00016-1

7.2 K-means

K-means is a widely utilized clustering technique known for its effectiveness in grouping data. This approach is characterized by its iterative, numerical, non-deterministic, and unsupervised nature. K-means has demonstrated successful applications in various research areas, particularly in the field of classification [1], [2].

K-means generates groups, a mean value of elements in the group represents each group. A collection of n elements is divided into k classes to minimize the similarity between clusters and maximize the similarity among elements within each cluster. The measure of resemblance is computed using the Euclidean distance.

Typically, the K-means algorithm categorizes elements into K clusters $C = c_1, c_2, \ldots, c_k$, where each C is associated with a clustering center μ_k. Subsequently, the Euclidean distance formula is employed to compute the sum of squared distances between the elements x_i within the cluster and the cluster center μ_k.

K-means minimizes the mean of the squared sum of the distance:

$$mean(c_k) = \sum_{x_i \in c_k} x_i - \mu_k \tag{7.1}$$

$$mean(K_c) = \sum_{k=1}^{K} mean(c_k) \tag{7.2}$$

7.3 Evaluation criteria

The classification capacity of the different models is evaluated by different criteria. One of the most used is the Receiver Operating Characteristic (ROC) curve, which compares the rate of TP true positives and FP false positives that vary between other cutoff points.

If true positives and false positives are equal, a line with poor or random classification performance is shown on the ROC curve. Thus the higher the line of the ROC curve, the better the performance shown by the classification method. If the classifier separates all the TP cases from the FP cases, the curve passes through the points $TP = 1$ and $FP = 0$ [3].

The summary generated by the ROC curve is called the area under the curve (AUC). The closer the AUC is to one, the better the ability of a classification method to distinguish between classes.

A detailed summary of the classification results of the models is provided by the confusion matrix. In addition to considering TP and FP, the confusion matrix considers cases correctly classified as negative class TN and cases incorrectly classified as negative class FN. This information obtained from the confusion matrix is used to evaluate the neural models with the following criteria:

Classification accuracy (*CA*) indicates the relationship between the number of correct predictions concerning the total number of samples:

$$CA = \frac{TP + FP}{TP + TN + FP + FN} \tag{7.3}$$

Precision (P) shows the classification performance of the DNNs of the true positives with respect to the false positives:

$$P = \frac{TP}{TP + FP} \tag{7.4}$$

The *Recall (R)* provides information on the number of true positives correctly identified:

$$R = \frac{TP}{TP + FN} \tag{7.5}$$

Finally, the score of *F*1 is helpful when the distribution of classes is uneven. This metric is obtained by combining *P* and *R* to compare the combined performance of *Precision* and *Recall* between various solutions:

$$F1 = 2 * \frac{P * R}{P + R} \tag{7.6}$$

7.4 Results

The results obtained by the neural classifiers for the classes of diabetes patients, patients with glucose intolerance or healthy patients are presented in this section.

Typically, neural networks use 70% of the data and 30% for testing to avoid overfitting. Then, it is said that the model avoids overfitting when the data obtained between training data and test data are not very different from each other. Furthermore, the neural models were validated with unknown data (data not used in the training or testing stages).

The resulting efficiency of the neural models according to the evaluation criteria is shown below for each of the networks used. It should be noted that deep neural networks were designed mainly to handle images, however, they were modified to manipulate and classify time series.

7.4.1 MLP results

7.4.1.1 Virtual patient classification results

The results obtained from the ROC curve of simulated patients from the MLP neuronal classifier are shown in Fig. 7.1. According to the ROC curve, the MLP model has an average curve of 0.99. For the class of normal patients, the area under the curve was 0.99, similar to the type of patients with glucose intolerance and diabetes, i.e., 0.99. In addition to the MLP performance according to the evaluation criteria described above, it is summarized in Table 7.1. The results of the different criteria were obtained from the confusion matrix shown in Fig. 7.2.

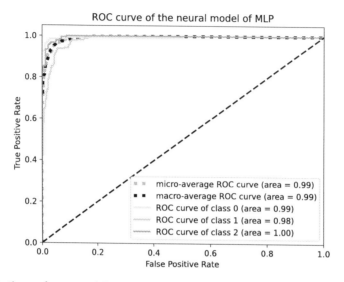

FIGURE 7.1 Classification performance of the MLP model. Class 0: normal patient, class 1: patient with impaired glucose tolerance, and class 2: patient with diabetes mellitus.

Table 7.1 Performance results of MLP model under the evaluation criteria described above.

Model	Class	AUC	CA	P(%)	R(%)	F1-Score
	Normal	0.99	0.92748	0.9489	0.96296	0.95588
MLP	IGT	0.98	0.92748	0.89455	0.95636	0.92442
	Diabetes Mellitus	1	0.92748	0.96913	0.85792	0.91014

According to the accuracy shown by the MLP neuronal classifier, it shows a classification performance of 0.92. Another relevant criterion is shown by the F1-score, which takes into consideration the precision and recall. The MLP receives an F1-score between 0.91 and 0.95 for the three classes. The results of the MLP are acceptable, but not good enough.

7.4.1.2 Real patient classification results

Table 7.3 presents a summary of the results for the different evaluation criteria obtained from the confusion matrix. Table 7.2 shows the classes obtained from the classification of the real data.

The classification results of the MLP are quite promising because it has a high percentage of classification accuracy over true classes. The MLP manages to detect the diabetic class without any problem, followed by the normal class. The most difficult class to diagnose belongs to the glucose intolerant patient.

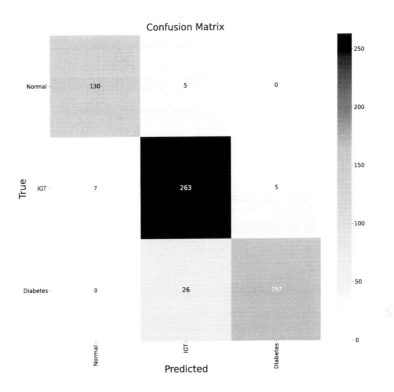

FIGURE 7.2 Confusion plot for MLP model.

Table 7.2 MLP classification results of real patients. N indicates the category of average patient, IGT indicates impaired glucose tolerance, and DM represents a diabetic patient.

Patient	a)	b)	c)	d)	e)	f)	g)	h)	i)	j)	k)	l)	m)	n)	o)	p)
Diagnosis	N	N	N	DM	IGT	IGT	N	IGT	N	N	N	N	N	N	N	IGT

Table 7.3 MLP performance results for real patients.

Model	Class	CA	P(%)	R(%)	F1-Score
	Normal	0.9375	0.9	1	0.9473
MLP	IGT	0.9375	1	0.83333	0.9090
	Diabetes Mellitus	0.9375	1	1	1

7.4.2 CNN results

7.4.2.1 Virtual patient classification results

The results obtained from the ROC curve of simulated patients from the CNN neuronal classifier are shown in Fig. 7.3. According to the ROC curve, the CNN model has an average curve of 0.97. For the class of normal patients, the area under the curve was 0.98, similar to the type of patients with impaired glucose tolerance, i.e., 0.96 and diabetes, i.e., 0.95.

In addition to the CNN performance according to the evaluation criteria described above, it is summarized in Table 7.4. The results of the different criteria were obtained from the confusion matrix shown in Fig. 7.4.

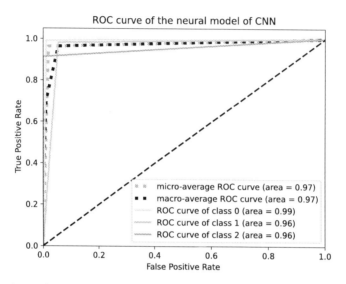

FIGURE 7.3 Classification performance of the CNN model. Class 0: normal patient, class 1: patient with impaired glucose tolerance, and class 2: patient with diabetes mellitus.

Table 7.4   Performance results of CNN model under the evaluation criteria described above.

Model	Class	AUC	CA	P(%)	R(%)	F1-Score
	Normal	**0.99**	0.95784	1	**0.96296**	0.98113
CNN	IGT	0.96	0.95784	0.91666	**1**	0.95652
	Diabetes Mellitus	0.96	0.95784	**1**	0.89071	0.94219

According to the accuracy shown by the CNN neuronal classifier, it shows a classification performance of 0.95. Another relevant criterion is shown by the F1-score, which takes into consideration the precision and recall. The CNN obtains an F1-score between 0.94 and 0.98 for the three classes. The CNN presents good results for classification of virtual patient data. The tests on real data are shown below to determine the generalization achieved by the network.

7.4.2.2 Real patient classification results
Table 7.6 presents a summary of the results for the different evaluation criteria obtained from the confusion matrix. Table 7.5 shows the classes obtained from the classification of the real data.

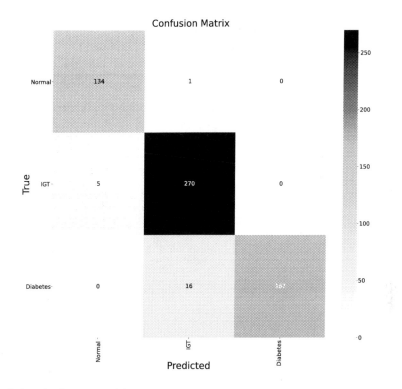

FIGURE 7.4 Confusion plot for CNN model.

Table 7.5   CNN classification results of real patients. N indicates the category of average patient, IGT indicates impaired glucose tolerance, and DM represents a diabetic patient.

Patient	a)	b)	c)	d)	e)	f)	g)	h)	i)	j)	k)	l)	m)	n)	o)	p)
Diagnosis	N	IGT	N	DM	N	IGT	N	IGT	DM	N	N	IGT	N	N	IGT	IGT

Table 7.6   CNN performance results for real patients.

Model	Class	CA	P(%)	R(%)	F1-Score
	Normal	0.8125	1	0.77777	0.875
CNN	IGT	0.8125	0.71428	0.83333	0.76923
	Diabetes Mellitus	0.8125	0.5	1	0.66666

The results shown by CNN for classifying data are not good enough. It has an accuracy of 0.81. CNN classifies the class of real patients better than the rest. The worst results are in the detection of the type of diabetic patient. In general. The CNN must be trained with more real data to improve its performance.

7.4.3 LSTM results

7.4.3.1 Virtual patient classification results

The results obtained from the ROC curve of simulated patients from the LSTM neuronal classifier are shown in Fig. 7.5. According to the ROC curve, the LSTM model has an average curve of 0.98. For the class of normal patients, the area under the curve was 0.96, similar to the type of patients with glucose intolerance, i.e., 0.97, and diabetes, i.e., 0.99. In addition to the LSTM performance according to the evaluation criteria described above, it is summarized in Table 7.7. The results of the different criteria were obtained from the confusion matrix shown in Fig. 7.6.

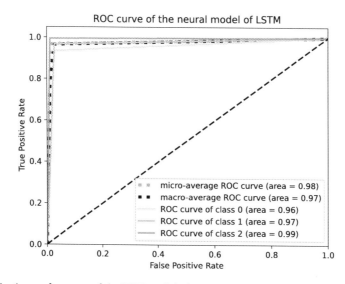

FIGURE 7.5 Classification performance of the LSTM model. Class 0: normal patient, class 1: patient with impaired glucose tolerance, and class 2: patient with diabetes mellitus.

Table 7.7 Performance results of LSTM model under the evaluation criteria described above.

Model	Class	AUC	CA	P(%)	R(%)	F1-Score
	Normal	0.96	0.96458	0.96153	0.92592	0.94339
LSTM	IGT	0.97	0.96458	0.96014	0.96363	0.96188
	Diabetes Mellitus	0.99	0.96458	0.97319	**0.99454**	**0.98369**

According to the accuracy shown by the LSTM neuronal classifier, it shows a classification performance of 0.96. Another relevant criterion is shown by the F1-score, which takes into consideration the precision and recall. The LSTM receives an F1-score between 0.94 and 0.98 for the three classes. The LSTM presents good results for classification of virtual patient data.

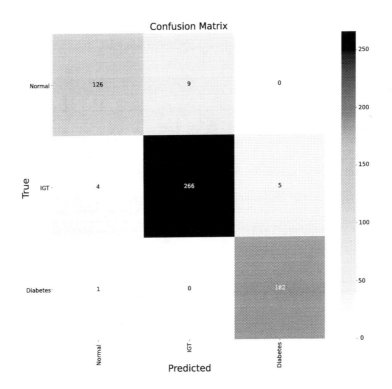

FIGURE 7.6 Confusion plot for LSTM model.

7.4.3.2 Real patient classification results

Table 7.9 presents a summary of the results for the different evaluation criteria obtained from the confusion matrix. Table 7.8 shows the classes obtained from the classification of the real data.

Table 7.8 LSTM classification results of real patients. N indicates the category of average patient, IGT indicates impaired glucose tolerance, and DM represents a diabetic patient.

Patient	a)	b)	c)	d)	e)	f)	g)	h)	i)	j)	k)	l)	m)	n)	o)	p)
Diagnosis	N	IGT	N	DM	N	N	N	IGT	IGT	N	N	IGT	N	N	N	IGT

Table 7.9   LSTM performance results for real patients.

Model	Class	CA	P(%)	R(%)	F1-Score
	Normal	0.875	0.88888	0.88888	**0.88888**
LSTM	IGT	0.875	0.83333	0.83333	**0.83333**
	Diabetes Mellitus	0.875	1	0.99453	**1**

The results shown by the LSTM to classify data are good, despite presenting problems in classifying the class of patients with glucose intolerance. It has an accuracy of 0.875. The LSTM is good at detecting patients with diabetes mellitus, obtaining an F1-score value of 1.

7.4.4 LSTM-FCN results

7.4.4.1 Virtual patient classification results

The results obtained from the ROC curve of simulated patients from the LSTM-FCN neuronal classifier are shown in Fig. 7.7. According to the ROC curve, the LSTM-FCN model has an average curve of 1. For the class of normal patients, the area under the curve was 1, similar to the type of patients with glucose intolerance and diabetes. In addition to the LSTM-FCN performance according to the evaluation criteria described above, it is summarized in Table 7.10. The results of the different criteria were obtained from the confusion matrix shown in Fig. 7.8.

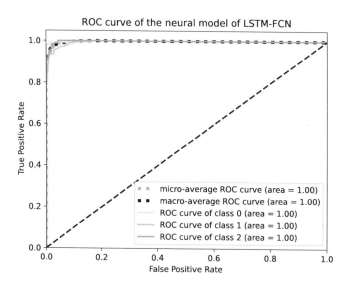

FIGURE 7.7 Classification performance of the LSTM-FCN model. Class 0: normal patient, class 1: patient with impaired glucose tolerance, and class 2: patient with diabetes mellitus.

Table 7.10 Performance results of LSTM-FCN model under the evaluation criteria described above.

Model	Class	AUC	CA	P(%)	R(%)	F1-Score
	Normal	1	0.91736	**0.97058**	0.97777	0.97416
LSTM-FCN	IGT	1	0.91736	0.85987	0.98181	0.91680
	Diabetes Mellitus	1	0.91736	0.9930	0.77595	0.87116

According to the accuracy shown by the LSTM-FCN neuronal classifier, it shows a classification performance of 0.91. Another relevant criterion is shown by the F1-score, which

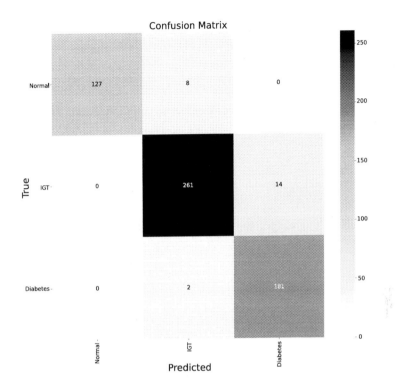

FIGURE 7.8 Confusion plot for LSTM-FCN model.

takes into consideration the precision and recall. The LSTM-FCN obtains an F1-score be-tween 0.87 and 0.97 for the three classes. The LSTM-FCN presents good results for classifi-cation of virtual patient data.

7.4.4.2 Real patient classification results

Table 7.12 presents a summary of the results for the different evaluation criteria obtained from the confusion matrix. Table 7.11 shows the classes obtained from the classification of the real data.

Table 7.11   LSTM-FCN classification results of real patients. N indicates the category of average patient, IGT indicates impaired glucose tolerance, and DM represents a diabetic patient.

Patient	a)	b)	c)	d)	e)	f)	g)	h)	i)	j)	k)	l)	m)	n)	o)	p)
Diagnosis	N	IGT	N	DM	DM	DM	N	DM	DM	N	N	DM	N	N	DM	IGT

The LSTM-FCN presents problems in classifying real data, despite showing excellent results in classifying virtual patient data.

Table 7.12   LSTM-FCN performance results for real patients.

Model	Class	CA	P(%)	R(%)	F1-Score
	Normal	0.625	0.85714	0.66666	0.75
LSTM-FCN	IGT	0.625	0.6	0.5	0.54545
	Diabetes Mellitus	0.625	0.25	1	0.4

7.4.5 ResNet results

7.4.5.1 Virtual patient classification results

The results obtained from the ROC curve of simulated patients from the ResNet neuronal classifier are shown in Fig. 7.9. According to the ROC curve, the ResNet model has an average curve of approximately 0.83. For the class of normal patients, the area under the curve was 0.85, similar to the type of patients with glucose intolerance and diabetes, i.e., 0.79. In addition to the ResNet performance according to the evaluation criteria described above, it is summarized in Table 7.13. The results of the different criteria were obtained from the confusion matrix shown in Fig. 7.10.

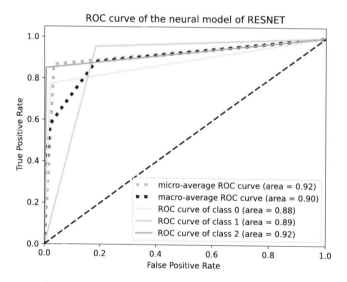

FIGURE 7.9 Classification performance of the ResNet model. Class 0: normal patient, class 1: patient with impaired glucose tolerance, and class 2: patient with diabetes mellitus.

Table 7.13   Performance results of ResNet model under the evaluation criteria described above.

Model	Class	AUC	CA	P(%)	R(%)	F1-Score
	Normal	0.88	0.87015	0.91603	0.88888	0.90225
ResNet	IGT	0.89	0.87015	0.93043	0.77818	0.84752
	Diabetes Mellitus	0.92	0.87015	0.78448	99453	0.87710

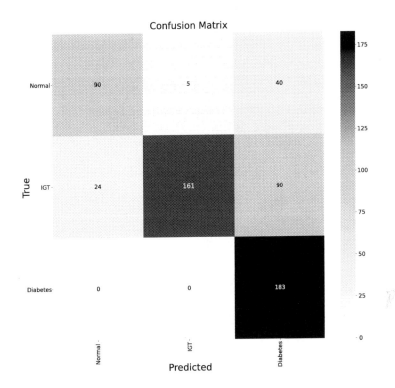

FIGURE 7.10 Confusion plot for ResNet model.

According to the accuracy shown by the ResNet neuronal classifier, it shows a classification performance of 0.87. Another relevant criterion is shown by the F1-score, which takes into consideration the precision and recall. The ResNet gets an F1-score between 0.84 and 0.90 for the three classes. The ResNet presents good results for classification of virtual patient data.

7.4.5.2 Real patient classification results

Table 7.15 presents a summary of the results for the different evaluation criteria obtained from the confusion matrix. Table 7.14 shows the classes obtained from the classification of the real data.

Table 7.14 ResNet classification results of real patients. N indicates the category of average patient, IGT indicates impaired glucose tolerance, and DM represents a diabetic patient.

Patient	a)	b)	c)	d)	e)	f)	g)	h)	i)	j)	k)	l)	m)	n)	o)	p)
Diagnosis	N	IGT	N	DM	N	N	N	IGT	DM	N	N	IGT	N	N	IGT	IGT

ResNet has complications in classifying patients with diabetes mellitus. However, the classification performance of normal and impaired glucose tolerance patients is good,

Table 7.15　ResNet performance results for real patients.

Model	Class	CA	P(%)	R(%)	F1-Score
	Normal	0.8125	1	0.77777	0.875
ResNet	IGT	0.8125	0.83333	0.83333	0.83333
	Diabetes Mellitus	0.8125	0.33333	1	0.5

which implies that the ResNet neural network should be trained with more examples of patients with diabetes.

7.4.6　K-Means results

7.4.6.1　Virtual patient classification results

The results obtained from the ROC curve of simulated patients from the K-Means neuronal classifier are shown in Fig. 7.11. According to the ROC curve, the K-Means model has an average curve of approximately 0.94. For the class of normal patients, the area under the curve was 0.93, similar to the type of patients with glucose intolerance and diabetes, i.e., 0.96. In addition to the K-Means performance according to the evaluation criteria described above, it is summarized in Table 7.16. The results of the different criteria were obtained from the confusion matrix shown in Fig. 7.12.

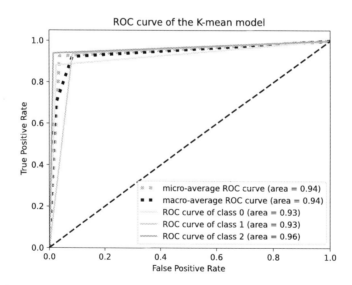

FIGURE 7.11 Classification performance of the K-Means model. Class 0: normal patient, class 1: patient with impaired glucose tolerance, and class 2: patient with diabetes mellitus.

According to the accuracy shown by the K-Means neuronal classifier, it shows a classification performance of 0.92. Another relevant criterion is shown by the F1-score, which takes into consideration the precision and recall. The K-Means gets an F1-score between

Table 7.16 Performance results of K-Means model under the evalua-tion criteria described above.

Model	Class	AUC	CA	P(%)	R(%)	F1-Score
	Normal	0.93	0.925	0.88095	0.90243	0.89156
K-Means	IGT	0.93	0.925	0.93548	0.90625	0.92063
	Diabetes Mellitus	0.96	0.925	0.93846	0.96825	0.95312

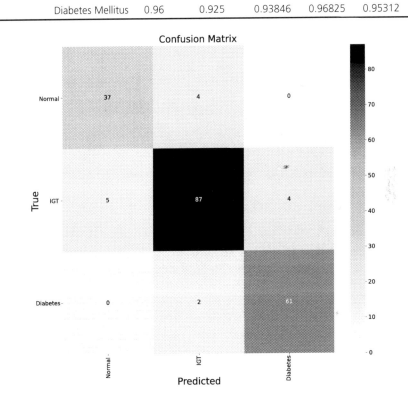

FIGURE 7.12 Confusion plot for K-Means model.

0.89 and 0.95 for the three classes. This means that the K-Means algorithm is good at clas-sifying virtual diabetes mellitus patient data.

7.4.6.2 Real patient classification results

Table 7.18 presents a summary of the results for the different evaluation criteria obtained from the confusion matrix. Table 7.17 shows the classes obtained from the classification of the real data.

K-Means has good classification performance for the diabetic patient class, and also presents good results for the classification of the remaining two classes (Normal and im-paired glucose tolerance). K-Means is good at generalizing data as it shows similar results for virtual and real patient data.

Table 7.17   K-Means classification results of real patients. N indicates the category of average patient, IGT indicates impaired glucose tolerance, and DM represents a diabetic patient.

Patient	a)	b)	c)	d)	e)	f)	g)	h)	i)	j)	k)	l)	m)	n)	o)	p)
Diagnosis	N	IGT	N	DM	N	IGT	N	IGT	IGT	N	N	IGT	N	N	N	IGT

Table 7.18   K-Means performance results for real patients.

Model	Class	CA	P(%)	R(%)	F1-Score
	Normal	0.875	0.8888	0.8888	0.8888
K-Means	IGT	0.875	0.8333	0.8333	0.8333
	Diabetes Mellitus	0.875	1	1	1

7.5 Results discussion

Due to the scarcity of information on patients with diabetes mellitus, glucose intolerance, and healthy subjects for training deep neural networks, the Dalla Man and UVA/Padova models were used. These models were used to generate sequential data by varying more sensitive parameters, specifically in the glucose tolerance test. From this process, data sets were obtained that included 441 healthy individuals, 865 with glucose intolerance, and 668 with diabetes, which were used in both the training and testing phases. Validation of the models was carried out using sequential data from patients who underwent glucose tolerance testing.

The results revealed superior performance compared to a traditional unsupervised machine learning method such as K-means algorithm. Although the random forest model presented favorable results on virtual data, its accuracy on real data reached around 93% with an F1 score between 0.8 and 1. In contrast, the LSTM model demonstrated slightly higher results, achieving an accuracy of 96% and an F1 score between 0.9 and 1 in real data. Consequently, it was concluded that the LSTM model has a more effective generalization ability than the random forest.

Additionally, the findings highlighted the fact that deep neural networks are valuable tools for disease classification and diagnosis, with an accuracy close to 96% in most models. Among these networks, LSTM stood out by showing the best results in terms of CA score, Recall (R), and F1 score metrics on real data. Although the performance of the other networks was similar, it was observed that the LSTM exhibited a higher area under the curve (AUC) compared to the other architectures.

References

[1] Sergio Mourelo Ferrandez, Timothy Harbison, Troy Weber, Robert Sturges, Robert Rich, Optimization of a truck-drone in tandem delivery network using K-means and genetic algorithm, Journal of Industrial Engineering and Management 9 (2) (2016) 374–388.

[2] Pengfei Shan, Image segmentation method based on K-mean algorithm, EURASIP Journal of Image and Video Processing 2018 (1) (2018) 1–9.

[3] Jaime Cerda, Lorena Cifuentes, Uso de curvas roc en investigación clínica: Aspectos teórico-prácticos, Revista Chilena de Infectología 29 (2) (2012) 138–141.

8

Conclusions and future work

8.1 Conclusions of Chapter 1

Science stands on the shoulders of giants, forged by the vast technology that the brilliant human mind has given us. In the last year, we have observed the takeoff of artificial intelligence, a computer system programmed to automate, predict, and create results from previously loaded data with the ability to feed back. These characteristics make it the most useful tool of the present century. However, there is still a long way to go to develop them adequately to a level at which they can completely resolve the biological problems that Diabetes Mellitus represents. Diabetes Mellitus is a slow, silent, and potentially fatal disease, which deteriorates the quality of life of people who suffer from it to such an extent that they require drastically changing all their activities; mainly diet, physical activity, recreational activities, family, work and social relationships, and even in personal care. The high complexity of its pathophysiological mechanisms makes it extremely difficult to predict its appearance beyond the risk present in multiple socio-environmental and other biological factors, without being able to identify the turning point to stop its appearance. Despite all the efforts in medicine to identify new biomarkers, the cornerstone of Diabetes continues to be the increase in serum glucose and its inability to be reduced through endogenous mechanisms. If adequate control is not maintained, both of the disease and of daily living habits, the injuries caused by Diabetes will increase until the existence of human life is exhausted, not only as an individual, but as a society. Following the treatment goals is essential, but it is even more important to eliminate modifiable predisposing factors and address those that cannot be eliminated to reduce their risk of appearance.

This is why the CGM is currently the best measurable parameter in real time to identify slight fluctuations and find the patterns necessary to predict physiological changes in time.

We have faith that technology advances sufficiently to obtain a reliable system, based on the pathophysiological aspects identified by the OGTT and scale it to a CGM, that is easy to feed, is adaptable, and provides tangible data that we can translate into effective measures that provide Public Health with a tool to identify the appearance of Diabetes in the early stages of the disease as well as predict the onset of its complications.

8.2 Conclusions of Chapter 2

The challenges and problems of diabetes mellitus were established. One of the most relevant topics that scientists have shown an interest in is the artificial pancreas, the objective of which is to provide a better quality of life to patients with diabetes mellitus. This device consists of a subcutaneous insulin system, a continuous glucose sensor (CGM), and a control algorithm. One of the main challenges of this device is the control, since it is respon-

Bio-Inspired Strategies for Modeling and Detection in Diabetes Mellitus Treatment
https://doi.org/10.1016/B978-0-44-322341-9.00017-3

sible for calculating insulin doses according to the patient's specific conditions. However, for the synthesis of the controller, mathematical models are necessary that faithfully describe the dynamics of glucose. To do so, external dynamics such as stress or exercise must be captured. The models must address this complexity to be effective. Various works have focused on the development of techniques for modeling, identification, and prediction of blood glucose levels.

Modeling techniques, ranging from mathematical models to machine learning techniques, have been applied to understand and represent the dynamics of diabetes. The approach depends on the availability of data in addition to the objective of the research.

The construction of models has also been developed from a predictive point of view. In the case of diabetes, it involves the use of historical data to predict future glucose levels and prevent risk scenarios such as hyperglycemia and hypoglycemia.

Effective diabetes prediction may benefit from personalized approaches that consider the unique characteristics of each individual. Accuracy of models is crucial for preventive interventions and effective disease management.

On the other hand, artificial intelligence models in the literature talk about the promising capacity for early detection and accurate classification of various diseases such as diabetes mellitus. These approaches can help to improve the accuracy of traditional techniques.

Furthermore, accurate classification of subtypes of diseases such as diabetes can help personalize treatment approaches, allowing for more effective management tailored to the individual needs of patients.

The detection and classification of diabetes mellitus represents great opportunities to improve medical care and its control. However, it also implies great ethical and technical challenges for the widespread acceptance of these technologies.

Modeling, identification, and prediction of diabetes mellitus are important and multidisciplinary areas that require a comprehensive approach, from data collection to clinical application. The integration of advanced modeling techniques and a patient-centered approach can contribute significantly to the understanding and effective management of this disease. In this chapter some of the mathematical models that describe the dynamics of glucose were shown in addition to experimental data obtained by a glucose sensor, an insulin pump, in addition to the glucose tolerance test.

8.3 Conclusions of Chapter 3

Evolutionary algorithms are versatile tools for solving complex optimization problems in various disciplines. Evolutionary algorithms are inspired by natural evolution and are useful when the optimal solution does not necessarily have to be found but an approximation is sufficient. These are useful for complex and multidimensional problems, where an exhaustive search would be impractical. Exploration and exploitation allow evolutionary algorithms to find global solutions close to optimal ones, especially in nonlinear and non-conventional problems. In addition, evolutionary algorithms are robust to noise in the data and can handle noisy fitness functions.

For their part, algorithms based on particle swarms have been used in optimization problems in different areas because they achieve an adequate balance between exploration of the search space and the exploitation of feasible regions, facilitating convergence towards optimal solutions. These algorithms are conceptually simple and elegant, allowing easy understanding and implementation compared to other more complex optimization approaches. These algorithms are often useful in high-dimensional problems, where traditional methods can become ineffective.

Neural networks have the ability to learn complex patterns as well as nonlinear representations from data, which allows them to be especially useful in tasks of pattern recognition, image processing, natural language processing, identification, among others. Neural networks can generalize from data sets in training to new data due to the various architectures of neural networks, such as convolutional neural networks (CNN) for images, recurrent neural networks (RNN) for temporal sequences, and more architectures such as deep neural networks (DNN), which allow us to address even more challenging problems.

In general, neural networks require large amounts of information in training to achieve good performance, which could be a limitation in applications where data is difficult to obtain, so new strategies for data extraction are necessary, mainly in the stage as crucial as training. Research on neural networks is constantly changing with new approaches and emerging architectures, making it a dynamic and rapidly growing field. Although they face challenges such as interpretability and the need for large amounts of data, their versatility and success in a wide range of applications make them a key technology in today's computing landscape.

Deep learning or deep neural networks have revolutionized the field of artificial intelligence and machine learning due to their learning capacity to capture abstraction patterns in high-dimensional data, thus demonstrating exceptional performance in various applications. Despite the success they have achieved in recent years, there are some limitations such as the large amount of data they need for their training, in addition to the Specialized Hardware necessary to carry it out. Overall, the potential of deep neural networks to boost innovation in areas such as medicine, autonomous driving or economics suggest a significant impact on society and technology. Deep neural networks are transforming the way complex problems in artificial intelligence and machine learning are being approached.

8.4 Conclusions of Chapter 4

Estimation of parameters for research and modeling of the glucose–insulin dynamics is essential in the field of medicine and treatment of diabetes, since descriptive models of this disease play a crucial role in blood sugar control, being fundamental in the development of glucose control algorithms. The construction of mathematical models that describe glucose–insulin dynamics allow a deeper understanding of the physiological processes involved, thus being valuable tools for the investigation of treatment strategies. These models commonly contain numerous parameters that involve adjusting parameter values based on observed data. This process is essential for the personalization of the models, improving the precision of the model and personalizing the treatment. The variability between individuals in terms of insulin response and glucose sensitivity highlights the need to obtain specific models. Parameter estimation allows models to be adapted to the specific characteristics of each patient. If an accurate estimation of the parameters of these models is achieved, it could contribute to the optimization of treatments, allowing finer adjustments in insulin doses and improving glycemic control. Among the main challenges, we can mention the variability of glucose dynamics from one day to the next or throughout the day due to external factors such as diet and physical activity that present difficulties in the accurate estimation of parameters.

The integration of mathematical models into emerging technologies such as continuous glucose monitors and automated insulin delivery devices opens the door to less invasive and more effective approaches to diabetes management. However, these models must first go through clinical validation and treatment protocols, so collaboration between scientists, engineers, and health professionals is essential to guarantee the effectiveness and safety of the generated strategies. The combination of mathematical models, parameter estimation techniques, and emerging technologies has the potential to transform the way we approach and treat metabolism-related diseases.

8.5 Conclusions of Chapter 5

Modeling glucose–insulin dynamics presents significant advances in the field of medicine and the treatment of diseases, especially diabetes. Neural networks offer to capture complex and nonlinear relationships of diabetes, which allows to more accurately capture the interactions between physiological variables. In addition, neural networks are especially useful when it comes to capturing dynamics from time series, that is, from data obtained by sensors, which is essential to understanding the dynamics over time of glucose metabolism and its insulin response. These features of neural networks that will capture complex patterns are shown to be superior to what traditional approaches have achieved. Furthermore, an identification approach with neural networks will open the door to the personalization of medicine, that is, adapting the models to the individual variables of the patients, which is fundamental in the treatment of diabetes where insulin resistance can vary significantly between patients. Among the advantages that are presented with respect to mathematical models in the literature, we can mention the integration of external factors such as physical activity, eating patterns, and stress that are not necessarily considered in most models, as two of the most recognized in diabetes research such as the Sorensen or Dalla Man model. Neural networks applied to diabetes modeling present challenges such as validation to guarantee confidence and effectiveness in real-world situations. The potential shown by modeling glucose–insulin dynamics in research can drive advances in the personalized management of diabetes by improving the precision and effectiveness of treatment.

8.6 Conclusions of Chapter 6

Multistep future prediction of blood glucose levels using neural networks is extremely beneficial in the treatment of diabetes mellitus. This approach would allow the anticipation of glycemic events such as hypoglycemia and hyperglycemia that would allow early intervention and improvement in disease management. It should be noted that the models to predict glucose levels can be adapted to changes in the patient's lifestyle such as their diet or exercise in addition to other factors, which results in improving their usefulness in dynamic and changing situations. Future prediction provides the patient with advanced information about their condition, which encourages informed decision making about the management of their disease. Successful implementation of future predictors requires collaboration of clinicians and scientists for continuous monitoring on accuracy and adaptability to ensure clinical relevance and effectiveness of the model. Due to the rapid evolution of technology, scientific research, and the incorporation of advances in the field of neural networks allow improvements in the efficiency and applicability of prediction models.

8.7 Conclusions of Chapter 7

The classification of time series is not a simple task, but it has a wide range of medical applications. There is a need to detect diseases from sensor-generated data in a rapid, accurate, and low-cost manner. The classification and detection of diabetes mellitus using deep neural network techniques and models is a relatively new area in research and healthcare. Having the ability to classify and detect diseases at early stages is extremely essential for timely interventions and treatments to prevent future complications. The first works to present a diagnostic approach were applying Machine Learning techniques and classification models to analyze large data sets, identify patterns, and facilitate the accurate detection and classification of different diabetes conditions, adapting to the variability in the data, clinical features, biomarkers, and patient risk profiles.

Despite the successes obtained by Machine Learning techniques, these models face challenges in generalization to different populations and in the interpretation of model decisions, which highlights the importance of rigorous validation.

Continuous efforts in research and development allow us to improve the accuracy and applicability of detection models. Another alternative is from deep neural networks, which allow the capture of complex and nonlinear patterns in the data, which is essential for the accurate identification of metabolic pathologies such as diabetes mellitus and glucose intolerance. The use of deep neural networks tends to improve the accuracy in classification and detection that provides more reliable results compared to traditional techniques.

8.8 General conclusions

Chronic diseases such as diabetes mellitus require enormous efforts on the part of patients, medical specialists, and governments. Therefore, techniques are required for the detection and treatment with intelligent techniques that help improve the quality of life of people who suffer from diabetes mellitus. On the one hand, evolutionary and particle swarm algorithms are especially useful in optimization problems such as parametric adjustment, so they have valuable potential in the parametric estimation of descriptive models of the dynamics of diabetes mellitus, a necessary task in the validation of mathematical models. Furthermore, when mathematical models are overwhelmed by the number of variables that intervene in the glucose–insulin system, modeling techniques through neural networks can be a feasible option in the acquisition of the mathematical models necessary in the synthesis of controllers and simulation of particular conditions of this disease. In addition, the models acquired by neural networks can be used to predict future glucose values that help identify risk conditions such as hypoglycemia and hyperglycemia. Two of the most used techniques in this task are direct and recursive configurations, both strategies have advantages and disadvantages between them, however, depending on the application you can choose between one or the other as appropriate.

For its part, the classification and detection of type 1 diabetes mellitus and impaired glucose tolerance present a challenge in the early and timely diagnosis of this disease based on time series. However, bio-inspired strategies are a promising approach to this task, providing accurate classifications. Among the most promising deep neural networks we can mention the convolutional neural network and long and short-term memory, whose performances have been shown to be encouraging.

As future work, neural models can be used in novel and reliable controllers for the regulation of blood glucose, in addition to implementing glucose predictors in devices such as sensors for alerts of risk conditions. Finally, disease detection and classification approaches should be sought based on data obtained online and the development of these diseases should be constantly monitored.

A

Model parameters

A.1 Dalla Man nominal parameter values

Table A.1 Dalla Man nominal parameter values.

$V_g = 1.49$	$m_5 = 0.0526$	$k_i = 0.0066$	$b = 0.68$
$K_1 = 0.065$	$m_6 = 0.8118$	$k_{max} = 0.0465$	$c = 0.00023$
$k_2 = 0.079$	$HE_b = 0.112$	$k_{min} = 0.0076$	$d = 0.009$
$V_I = 0.04$	$k_{p1} = 2.7$	$k_{abs} = 0.023$	$F_{cns} = 1$
$m_1 = 0.379$	$k_{p2} = 0.0021$	$k_{gri} = 0.0465$	$V_{m0} = 0.034$
$m_2 = 0.673$	$k_{p3} = 0.009$	$f = 0.9$	$V_{mx} = 0.034$
$m_4 = 0.269$	$k_{p4} = 0.0786$	$a = 0.00016$	$k_{m0} = 471.13$
$P_{2u} = 0.084$	$kd = 0.0164$	$Rho = 0.57$	$K = 0.99$
$\alpha = 0.013$	$k_{a1} = 0.0018$	$\delta = 0.682$	$k_{h1} = 0.0164$
$\beta = 0.05$	$k_{a2} = 0.0182$	$\sigma = 1.093$	$k_{h2} = 0.0018$
$\gamma = 0.5$	$kH = 0.16$	$n = 0.15$	$k_{h3} = 0.0182$
$ke_1 = 0.0007$	$ke_2 = 269$	$\zeta = 0.009$	$H_b = 93$

A.2 Sorensen nominal parameter values

Table A.2   Hemodynamic parameters.

$p_1 = \dfrac{Q_B^G}{V_{BV}^G}$	$p_{11} = \dfrac{Q_G^G}{V_G^G}$	$p_{21} = \dfrac{1}{T_P^G}$	$p_{31} = \dfrac{V_{PI}}{T_P^I V_{PV}^I}$	$p_{41} = \dfrac{1}{\tau_I}$
$p_2 = \dfrac{V_{BI}}{V_{BV}^G T_B}$	$p_{12} = \dfrac{1}{V_G^G}$	$p_{22} = \dfrac{1}{V_{PI}}$	$p_{32} = \dfrac{Q_G^I}{V_G^I}$	$p_{42} = 1.21$
$p_3 = \dfrac{1}{T-B}$	$p_{13} = \dfrac{1}{V_L^G}$	$p_{23} = \dfrac{Q_B^I}{V_B^I}$	$p_{33} = \dfrac{1}{V_L^I}$	$p_{43} = 1.14$
$p_4 = \dfrac{1}{V_{BI}}$	$p_{14} = Q_A^G$	$p_{24} = \dfrac{1}{V_H^I}$	$p_{34} = Q_A^I$	$p_{44} = 1.66$
$p_5 = \dfrac{1}{V_H^G}$	$p_{15} = Q_G^G$	$p_{25} = Q_B^I$	$p_{35} = Q_G^I$	$p_{45} = 21.43$
$p_6 = Q_B^G$	$p_{16} = Q_L^G$	$p_{26} = Q_L^I$	$p_{36} = Q_L^I$	$p_{46} = 0.84$
$p_7 = Q_L^G$	$p_{17} = \dfrac{Q_K^G}{V_K^G}$	$p_{27} = Q_K^I$	$p_{37} = \dfrac{Q_K^I}{V_K^I}$	$p_{47} = \dfrac{1}{65}$
$p_8 = Q_K^G$	$p_{18} = \dfrac{1}{V_K^G}$	$p_{28} = Q_P^I$	$p_{38} = \dfrac{1}{V_K^I}$	$p_{48} = 2.7$
$p_9 = Q_P^G$	$p_{19} = \dfrac{Q_P^G}{Q_{PV}^G}$	$p_{29} = Q_H^I$	$p_{39} = \dfrac{1}{T_P^I}$	$p_{49} = 0.39$
$p_{10} = Q_H^G$	$p_{20} = \dfrac{V_{PI}}{T_P^G V_{PV}^G}$	$p_{30} = \dfrac{Q_P^I}{V_{PV}^I}$	$p_{40} = \dfrac{1}{V_I^P}$	$p_{50} = 1$
$p_{51} = 2$	$p_{52} = \dfrac{1}{\tau_{GC}}$	$p_{53} = 2$	$p_{54} = 0.55$	$p_{55} = 21.43$
$p_{56} = \dfrac{1}{V_G^{GC}}$	$p_{57} = \dfrac{1}{Q_{PI}}$	$p_{58} = \dfrac{T_{PI}}{V_{PII}}$		

Table A.3   Metabolic parameters.

$p_{59} = 155$	$p_{64} = 0.62$	$p_{69} = 20$	$p_{74} = 6.52$	$p_{79} = 2.93$	$p_{84} = 1.06$
$p_{60} = 2.7$	$p_{65} = 0.497$	$p_{70} = 71$	$p_{75} = 0.338$	$p_{80} = 2.10$	$p_{85} = 0.47$
$p_{61} = 0.39$	$p_{66} = 5.66$	$p_{71} = 0.11$	$p_{76} = 70$	$p_{81} = 4.18$	$p_{86} = 35.0$
$p_{62} = 1.42$	$p_{67} = 2.44$	$p_{72} = 460$	$p_{77} = 10$	$p_{82} = 1.31$	$p_{87} = 0.0098$
$p_{63} = 1.41$	$p_{68} = 1.48$	$p_{73} = 7.03$	$p_{78} = 20$	$p_{83} = 0.61$	$p_{88} = 5.82$
$p_{89} = 2.55$	$p_{90} = 9.12$				

Index